Python Geospatial Analysis Cookbook

60 recipes to work with topology, overlays, indoor routing, and web application analysis with Python

Michael Diener

PUBLISHING

BIRMINGHAM - MUMBAI

Python Geospatial Analysis Cookbook

First published: November 2015

Production reference: 1251115

Published by Packt Publishing Ltd.
Livery Place
35 Livery Street
Birmingham B3 2PB, UK.

ISBN 978-1-78355-507-9

www.packtpub.com

Credits

Author
Michael Diener

Reviewers
Jáchym Čepický

Richard Marsden

Commissioning Editor
Ashwin Nair

Acquisition Editor
Rebecca Youé

Content Development Editor
Athira Laji

Technical Editor
Chinmay S. Puranik

Copy Editor
Sonia Michelle Cheema

Project Coordinator
Bijal Patel

Proofreader
Safis Editing

Indexer
Hemangini Bari

Graphics
Disha Haria

Production Coordinator
Nilesh Mohite

Cover Work
Nilesh Mohite

About the Author

Michael Diener graduated from Simon Fraser University, British Columbia, Canada, in 2001 with a bachelor of science degree in geography. He began working in 1995 with Environment Canada as a GIS (Geographic Information Systems) analyst and has continued to work with GIS technologies ever since.

In 2008, he founded a company called GOMOGI that is focused on building web and mobile GIS application with open source tools. In 2011, the focus changed to indoor wayfinding and navigation solutions and building the indrz platform that Michael had envisioned.

From time to time, Michael also holds seminars for organizations wanting to explore or discover the possibilities of how GIS can increase productivity and help better answer spatial questions. He is also the creative head of new product development in his company. His technical skills include working with Python to solve a wide range of spatial problems on a daily basis. Through the years, he has developed many spatial applications with Python, including indrz and golfgis, which are two of the products built by his company, GOMOGI.

He is also lecturer of GIS at the Alpen Adria University, Klagenfurt, where he enjoys teaching students the wonderful powers of GIS and explaining how to solve spatial problems with open source GIS and Python.

I would like to dedicate this book to my family, beginning with my loving wife, Silvia, who encouraged and supported me through the writing of this book. I must also thank my three boys, Noah, Levi, and Jasper, for sacrificing many evenings and weekends of playtime for daddy's book time. My parents, Birte and Hartwig, also deserve a big thank you for always believing in me—thank you for everything.

About the Reviewers

Jáchym Čepický is an open source GIS software consultant, developer, and user.

Richard Marsden has over 15 years of professional software development experience. After starting in the fields of geophysics and oil exploration, he has spent the last 10 years running the Winwaed Software Technology LLC independent software vendor. Winwaed specializes in geospatial tools and applications, including web applications, and operates the http://www.mapping-tools.com website for tools and add-ins for geospatial products such as Caliper Maptitude and Microsoft MapPoint.

Richard has been a technical reviewer for *Python Geospatial Development* and *Python Geospatial Analysis Essentials* both by Erik Westra, and *Mastering Python Forensics* by Dr. Michael Spreitzenbarth and Dr. Johann Uhrmann all by Packt Publishing.

www.PacktPub.com

Support files, eBooks, discount offers, and more

For support files and downloads related to your book, please visit www.PacktPub.com.

Did you know that Packt offers eBook versions of every book published, with PDF and ePub files available? You can upgrade to the eBook version at www.PacktPub.com and as a print book customer, you are entitled to a discount on the eBook copy. Get in touch with us at service@packtpub.com for more details.

At www.PacktPub.com, you can also read a collection of free technical articles, sign up for a range of free newsletters and receive exclusive discounts and offers on Packt books and eBooks.

https://www2.packtpub.com/books/subscription/packtlib

Do you need instant solutions to your IT questions? PacktLib is Packt's online digital book library. Here, you can search, access, and read Packt's entire library of books.

Why Subscribe?

- ▶ Fully searchable across every book published by Packt
- ▶ Copy and paste, print, and bookmark content
- ▶ On demand and accessible via a web browser

Free Access for Packt account holders

If you have an account with Packt at www.PacktPub.com, you can use this to access PacktLib today and view 9 entirely free books. Simply use your login credentials for immediate access.

Table of Contents

Preface

Geospatial analysis is not special; it is just different when compared to other types of analysis such as financial market analysis. We work with geometry objects, such as lines, points, and polygons, and connect these geometries to attributes such as business data. We ask "where" question, such as "Where is the nearest pub?", "Where are all my customers located?", and "Where is my competition located?". The other location questions include, "Will this new building cast a shadow over the park?", "What is the shortest way to school?", "What is the safest way to school for my kids?", "Will this building block my view of the mountains?", and "Where is the optimal place to build my next store?". Identify the areas that fire trucks can reach from their station in 5 min, 10 min, or 20 min, and so on.

One thing all these questions have in common is the fact that you need to know where certain objects are located in order to answer them. Without the *spatial* component, you cannot answer such questions and this is what geospatial analysis is all about.

Geospatial features are laid over each other and patterns or trends are easily identified. This ability to see a pattern or trend is geospatial analysis in its simplest form.

Throughout this book, simple and complex code recipes are provided as small working models that can easily be integrated or expanded into a larger project or model.

Analysis is the fun part of GIS, and involves visualizing relationships, identifying trends, and seeing patterns that are not visible in a spreadsheet.

The Python programming language is clean, clear, and concise, making it great for beginners. It also has advanced powers for professionals to help them quickly code solutions to complex problems. Python makes visualization quick and easy for experts or beginners who work with geospatial data. It's that simple.

What this book covers

Chapter 1, *Setting Up Your Geospatial Python Environment*, explores setting up your computer to handle all software requirements in one go, such as pyproj, NumPy, and Shapely. All your development software needs to enable spatial analysis or geoprocessing on Windows and Linux are met in this chapter.

Chapter 2, *Working with Projections*, explains how to deal with spatial data that's projected or unprojected. You can learn and discover how to transform your data into a correct projection to prepare for an analysis.

Chapter 3, *Moving Spatial Data from One Format to Another*, explains how geospatial data comes in many different formats and also how messaging data from one format to another is a daily chore. In this chapter, you will find out about the most common data management tasks.

Chapter 4, *Working with PostGIS*, shows you how most of our geospatial data is stored in a spatial database and using, accessing, manipulating this data with Python is what this chapter is about.

Chapter 5, *Vector Analysis*, introduces a very common geospatial data format, that is, the vector data format. To execute analysis functions on vector data, we will explore patterns used to create new data by snapping, clipping, cutting, and overlaying vector datasets followed by determining the 3D ground distance and total elevation gain.

Chapter 6, *Overlay Analysis*, explains how to combine spatial data to create new data by using the process of overlaying two sets of data over each other.

Chapter 7, *Raster Analysis*, shows you how to create an elevation profile and quick ways to merge images to perform raster analysis functions on your data.

Chapter 8, *Network Routing Analysis*, shows you how finding the nearest anything is a common geospatial analysis feature. This chapter will disclose how to go about solving an indoor network type problem and demonstrate some common use cases for wayfinding inside buildings.

Chapter 9, *Topology Checking and Data Validation*, covers data quality and connections. In this chapter, you will learn how to verify your data for errors using custom topological functions.

Chapter 10, *Visualizing Your Analysis*, explains how geospatial data is inherently visual and you will learn about presenting your analysis on a web map and a 3D web.

Chapter 11, *Web Analysis with GeoDjango*, builds on *Chapter 8*, *Network Routing Analysis*, where you will create an indoor routing web application. You will easily be able to route a person from point A to point B within a building with real 3D network data. These key features will be presented by bringing together all the parts of the recipes you have learned so far.

Appendix A, Other Geospatial Python Libraries, explains how Python flourishes with geospatial libraries, and you will also find a listing of many popular libraries that are used for data analysis, regardless of whether they're spatial or not. This may trigger your interest.

Appendix B, Mapping Icon Libraries, quickly goes over the icon libraries out there that play a special role in the python geospatial working environment.

What you need for this book

To work with this book, you should be familiar with the programming language Python and the concepts involved in programming. This means that you should be able to install Python 2.7.x on your machine (Windows, Linux, or OS X) if it's not already installed. The concepts related to GIS (Geographic Information Systems) are definitely helpful but not necessary. A primer to this book could be *Learning Geospatial Analysis with Python, Joel Lawhead* or *Python Geospatial Development, Eric Westra*, both by *Packt Publishing*.

Who this book is for

If you are a student, teacher, programmer, geospatial or IT administrator, GIS analyst, researcher, or scientist looking to learn about spatial analysis, then this book is for you. Anyone trying to answer simple to complex spatial analysis questions will get a working demonstration of the power of Python with the help of real-world data. Some of you may be beginners but most of you will probably have a basic understanding of geospatial analysis and programming.

Sections

In this book, you will find several headings that appear frequently (Getting ready, How to do it..., How it works..., There's more..., and See also).

To give clear instructions on how to complete a recipe, we use these sections as follows:

Getting ready

This section tells you what to expect in the recipe, and describes how to set up any software or any preliminary settings required for the recipe.

How to do it...

This section contains the steps required to follow the recipe.

How it works...

This section usually consists of a detailed explanation of what happened in the previous section.

There's more...

This section consists of additional information about the recipe in order to make the reader more knowledgeable about the recipe.

See also

This section provides helpful links to other useful information for the recipe.

Conventions

In this book, you will find a number of styles of text that distinguish between different kinds of information. Here are some examples of these styles, and an explanation of their meaning.

Code words in text are shown as follows: "If `workon` for some reason does not start your virtual environment, you can start it simply by executing `source /home/mdiener/.venvs/ pygeoan_cb/bin/activate` from the command line."

A block of code is set as follows:

```python
#!/usr/bin/env python
# -*- coding: utf-8 -*-

from osgeo import ogr
shp_driver = ogr.GetDriverByName('ESRI Shapefile')
shp_dataset = shp_driver.Open(r'../geodata/schools.shp')
shp_layer = shp_dataset.GetLayer()
shp_srs = shp_layer.GetSpatialRef()
print shp_srs
```

Any command-line input or output is written as follows:

```
$ sudo apt-get install python-setuptools python-pip
```

New terms and **important words** are shown in bold. Words that you see on the screen, in menus or dialog boxes for example, appear in the text like this: "Select **Route To:** and enter 2 to see the second floor options."

 Warnings or important notes appear in a box like this.

Tips and tricks appear like this.

Reader feedback

Feedback from our readers is always welcome. Let us know what you think about this book— what you liked or may have disliked. Reader feedback is important for us to develop titles that you really get the most out of.

To send us general feedback, simply send an e-mail to `feedback@packtpub.com`, and mention the book title through the subject of your message.

If there is a topic that you have expertise in and you are interested in either writing or contributing to a book, see our author guide on `www.packtpub.com/authors`.

Customer support

Now that you are the proud owner of a Packt book, we have a number of things to help you to get the most from your purchase.

Downloading the example code

You can download the example code files for all Packt books you have purchased from your account at `http://www.packtpub.com`. If you purchased this book elsewhere, you can visit `http://www.packtpub.com/support` and register to have the files e-mailed directly to you.

Downloading the color images of this book

We also provide you with a PDF file that has color images of the screenshots/diagrams used in this book. The color images will help you better understand the changes in the output. You can download this file from `https://www.packtpub.com/sites/default/files/downloads/5079OS_ColorImage.pdf`.

Errata

Although we have taken every care to ensure the accuracy of our content, mistakes do happen. If you find a mistake in one of our books—maybe a mistake in the text or the code—we would be grateful if you would report this to us. By doing so, you can save other readers from frustration and help us improve subsequent versions of this book. If you find any errata, please report them by visiting http://www.packtpub.com/support, selecting your book, clicking on the **errata submission form** link, and entering the details of your errata. Once your errata are verified, your submission will be accepted and the errata will be uploaded to our website, or added to any list of existing errata, under the Errata section of that title.

Piracy

Piracy of copyright material on the Internet is an ongoing problem across all media. At Packt, we take the protection of our copyright and licenses very seriously. If you come across any illegal copies of our works, in any form, on the Internet, please provide us with the location address or website name immediately so that we can pursue a remedy.

Please contact us at copyright@packtpub.com with a link to the suspected pirated material.

We appreciate your help in protecting our authors, and our ability to bring you valuable content.

Questions

You can contact us at questions@packtpub.com if you are having a problem with any aspect of the book, and we will do our best to address it.

1
Setting Up Your Geospatial Python Environment

In this chapter, we will cover the following topics:

- ▶ Installing virtualenv and virtualenvwrapper
- ▶ Installing pyproj and NumPy
- ▶ Installing shapely, matplotlib, and descartes
- ▶ Installing pyshp, geojson, and pandas
- ▶ Installing SciPy, PySal, and IPython
- ▶ Installing GDAL and OGR
- ▶ Installing GeoDjango and PostgreSQL with PostGIS

Introduction

This chapter will get the grunt work done for you so that you can freely and actively complete all the recipes in this book. We will start off by installing, each of the libraries you will be using, one by one. Once each step is completed, we will test each library installation to make sure it works. Since this book is directed toward those of you already working with spatial data, you can skip this chapter if you have it installed already. If not, you will find the installation instructions here useful as a reference.

The choice of Python libraries is based on industry-proven reliability and functionality. The plethora of functions in Python libraries has led to a flourishing GIS support on many top desktop GIS systems, such as QGIS and ESRI ArcGIS.

Also included in this book is an `installer.sh` bash file. The `installer.sh` file can be used to install the Python libraries that are available for your virtual environment from `pip` and other dependencies via the `apt-get` command. The `installer.sh` bash file is executed from the command line and installs almost everything in one go, so please take a look at it. For those of you who are starting with Python for the first time, follow the instructions in this chapter and your machine will be set up to complete different recipes.

Installations can sometimes be tricky even for advanced users, so you will find some of the most common pitfalls and hook-ups described in this chapter.

The development of these recipes was completed on a fresh **Linux/Ubuntu 14.04** machine. Therefore, the code examples, if not otherwise specified, are Linux/Ubuntu-specific with Windows notes wherever necessary, unless otherwise specified.

Installing virtualenv and virtualenvwrapper

This recipe will enable you to manage different versions of different libraries for multiple projects. We use `virtualenv` to create virtual Python environments to host collections of project-specific libraries in an isolated directory. For example, you may have an old legacy project using Django 1.4, whereas a new project requires you use Django version 1.8. With `virtualenv`, you can have both versions of Django installed on the same machine, and each project can access the appropriate version of Django without any conflicts or problems.

Without `virtualenv`, you are forced to either upgrade the old project or find a workaround to implement the new features of the other version, therefore limiting or complicating the new project.

The `virtualenv` allows you to simply switch between different Python virtual environments for your individual projects. This has the added benefit that you can easily and quickly set up a new machine for testing or help a new developer get their machine up and running as fast as possible.

Getting ready

Before anything, we are going to assume that you already have a Linux/Ubuntu machine or a **virtualbox** instance running Linux/Ubuntu so you can follow these instructions.

 I also suggest trying out Vagrant (`http://www.vagrantup.com`), which uses virtualbox to box and standardize your development environment.

Ubuntu 14.04 comes with Python 2.7.6 and Python 3.4 preinstalled; the other libraries are your responsibility as explained in the following sections.

Windows users need to download and install Python 2.7.x from the Python home page at `https://www.python.org/downloads/windows/`; please download the newest version of the 2.7.x series since this book is written with 2.7.X in mind. The installer includes a bundled version of pip, so make sure you install it!

Take a close look at the correct version to download, making sure that you get either the *32-bit* or *64-bit* download. You cannot mix and match the versions, so be careful and remember to install the correct version.

A great site for other kinds of Windows binaries can be found at `http://www.lfd.uci.edu/~gohlke/pythonlibs/`. Wheel files are the new norms of installations and can be executed from the command line as follows:

```
python pip install libraryName.whl
```

> On Windows, make sure that your Python interpreter is set up on your system path. This way, you can simply call Python directly from the command prompt using the `C:\Users\Michael> python filename.py` command. If you need more help, information can be found by following one of the online instructions at `https://pypi.python.org/pypi/pip`.
>
> As of Python 2.7.9 and later versions, `pip` is available on installation.

Python 3 would be awesome to use, and for many Python GIS libraries, it is ready for show time. Unfortunately, not all GIS libraries jive with Python 3 (pyproj) as one would love to see at the time of writing this. If you want, feel free to go for Python 3.x and give it a go. A great webpage to check the compatibility of a library can be found at `https://caniusepython3.com/`.

To install `virtualenv`, you need to have a running installation of Python and pip. The pip package manager manages and installs Python packages, making our lives easier. Throughout this book, if we need to install a package, `pip` will be our tool of choice for this job. The official installation instructions for pip can be found at `https://pip.pypa.io/en/latest/installing.html`. To install pip from the command line, we first need to install `easy_install`. Let's try it out from the Terminal:

```
$ sudo apt-get install python-setuptools python-pip
```

With this one line, you have both `pip` and `easy_install` installed.

What is sudo?

sudo is a program for Unix-like computer operating systems that allows users to run programs with the security privileges of another user (normally, the super user or root). Its name is a concatenation of **su** (**substitute user**) and **do** (**take action**). Take a look at `http://en.wikipedia.org/wiki/Sudo` for more information on this.

The command **sudo** means to run an execution as a super user. If this fails, you will need to get the `ez_setup.py` file, which is available at `https://bootstrap.pypa.io/ez_setup.py`. After downloading the file, you can run it from the command line:

```
$ python ez_setup.py
```

Now `pip` should be up and running and you can execute commands to complete the installations of **virtualenv** and **virtualenvwrapper**. The `virtualenvwrapper` creates shortcuts that are faster ways to create or delete your virtual environments. You can test it as follows:

```
$ pip install virtualenv
```

How to do it...

The steps to install your Python `virtualenv` and `virtualenvwrapper` packages are as follows:

1. Install `virtualenv` using the pip installer:

   ```
   $ sudo pip install virtualenv
   ```

2. Install `virtualenvwrapper` using `easy_install`:

   ```
   $ sudo easy_install virtualenvwrapper
   ```

   We use `easy_install` instead of `pip` because with Ubuntu 14.04, the `virtualenvwrapper.sh` file is unfortunately not located at `/usr/local/bin/virtualenvwrapper.sh` where it should be according to the online documentation.

3. Assign the `WORKON_HOME` variable to your home directory with the folder name `venvs`. Create a single folder where you want to store all your different Python virtual environments; in my case, the folder is located at `/home/mdiener/venvs`:

   ```
   $ export WORKON_HOME=~/venvs
   $ mkdir $WORKON_HOME
   ```

4. Run the source command to execute the `virtualenvrapper.sh` bash file:

```
$ source /usr/local/bin/virtualenvwrapper.sh
```

5. Next, we create a new virtual environment called `pygeoan_cb`, and this is also the name of the new folder where the virtual environment is installed:

```
$ mkvirtualenv pygeoan_cb
```

To use `virtualenvwrapper` the next time you start up your machine, we need to set it up so that your bash terminal runs the `virtualenvwrapper.sh` script when your computer starts.

6. First, put it in your `~/.bashrc` file:

```
$ echo "export WORKON_HOME=$WORKON_HOME" >> ~/.bashrc
```

7. Next, we'll import the `virtualenvwrapper` function in our bash:

```
$ echo "source /usr/local/bin/virtualenvwrapper.sh" >> ~/.bashrc
```

8. Now we can execute our bash:

```
$ source ~/.bashrc
```

How it works...

Step one shows how pip installs the `virtualenv` package into your system-wide Python installation. Step two shows how the `virtualenvwrapper` helper package is installed with `easy_install` because the `virtualenvwrapper.sh` file is not created using the pip installer. This will help us create, enter, and generally, work or switch between Python virtual environments with ease. Step three assigns the WORKON_HOME variable to a directory where we want to have all of our virtual environments. Then, we'll create a new directory to hold all the virtual environments. In step four, the command source is used to execute the shell script to set up the `virtualenvwrapper` package. In step five, we see how to actually create a new `virtualenv` called `pygeoan_cb` in our `/home/mdiener/venvs` directory. This final step automatically starts our `virtualenv` session.

Once the `virtualenv` session starts, we can now see the name of `virtualenv` in brackets like this:

```
(pygeoan_cb)mdiener@mdiener-VirtualBox:~$
```

To exit `virtualenv`, simply type the following code:

```
$ deactivate
```

Now, your command line should be back to normal as shown here:

```
mdiener@mdiener-VirtualBox:~$
```

To reactivate `virtualenv`, simply type:

```
$ workon pygeoan_cb
```

 The `workon` command has *Tab* completion. So, simply type `workon`, and then the first letter of the name of the virtual environment you want to enter, such as `py`. Hit *Tab* and it will autocomplete the name.

Inside the `/venvs` folder, you will find specific individual virtual environments for each project in the form of a subfolder. The `virtualenvwrapper` package will always create a new folder for each new project you create. You can, therefore, easily delete a folder and it will remove your virtual environment.

To quickly print a list all of the installed libraries to a file, we'll use the `pip` command:

```
$ pip freeze > requirements.txt
```

This will create a text file called `requirements.txt` in the current folder. The text file contains a list of all the installed Python packages inside the Python virtual environment currently running.

To create a new `virtualenv` from a requirements file, use the following command:

```
$ pip install -r /path/to/requirements.txt
```

There's more...

For those of you who are just starting out with geospatial Python development, it should be noted that you should keep your project-specific code at another location outside your Python virtual environment folder. For example, I always have each project-related code contained in a separate folder called `01_projects`, which is my main folder. The path to my projects folder is `/home/mdiener/01_projects`, and the structure of two of my projects is as follows:

- ▸ `01_projects/Name_project1`
- ▸ `01_projects/Name_project2`

All virtual environments are located under `/home/mdiener/venvs/`. Usually, I give them the same name as a project to keep things organized, as follows:

- ▸ `/home/mdiener/venvs/Name_project1`
- ▸ `/home/mdiener/venvs/Name_project2`

Installing pyproj and NumPy

The **pyproj** is a wrapper around the PROJ.4 library that works with projections and performs transformations (https://pypi.python.org/pypi/pyproj/) in Python. All your geographic information should be projected into one of the many coordinate systems supported by the **European Petroleum Survey Group** (**EPSG**). This information is necessary for the systems to correctly place data at the appropriate location on Earth. The geographic data can then be placed on top of each other as layers upon layers of data in order to create maps or perform analysis. The data must be correctly positioned or we won't be able to add, combine, or compare it to other data sources spatially.

Data comes from many sources and, often, a projection is not the same as a dataset. Even worse, the data could be delivered with a description from a data provider stating it's in projection UTM31 when, in reality, the data is in projection UTM34! This can lead to big problems later on when trying to get your data to work together as programs will throw you some ugly error messages.

NumPy is the scientific backbone of number crunching arrays and complex numbers that are used to power several popular geospatial libraries including **GDAL** (**geospatial abstraction library**). The power of NumPy lies is in its support for large matrices, arrays, and math functions. The installation of NumPy is, therefore, necessary for the other libraries to function smoothly, but is seldom used directly in our quest for spatial analysis.

Getting ready

Fire up your virtual environment, if it is not already running, using the following standard start command:

```
$ workon pygeoan_cb
```

Your prompt should now look like this:

```
(pygeoan_cb)mdiener@mdiener-VirtualBox:~$
```

 If workon for some reason does not start your virtual environment, you can start it simply by executing source / home/mdiener/venvs/pygeoan_cb/bin/activate from the command line; try the steps listed in the *Installing virtualenv and virtualenvwrapper* recipe again to get it going.

Now, we need to install some Python tools for development that allow us to install NumPy, so run this command:

```
$ sudo apt-get install -y python-dev
```

You are now ready to move on and install pyproj and NumPy inside your running virtual environment.

How to do it...

Simply fire up `virtualenv` and we will use the pip installer to do all the heavy lifting as follows:

1. Use pip to go ahead and install NumPy; this can take a couple of minutes as many lines of installation verbosity are written on screen:

    ```
    $ pip install numpy
    ```

    Windows users can grab the `.whl` file for NumPy and execute it using following command:

    ```
    pip install numpy -1.9.2+mkl-cp27-none-win32.whl
    ```

2. Use `pip` one more time to install pyproj:

    ```
    $ pip install pyproj
    ```

    Windows users can use the following command to install pyproj:

    ```
    pip install pyproj-1.9.4-cp27-none-win_amd64.whl
    ```

3. Wait a few minutes; NumPy should be now running along with pyproj. To test if it's worked out, enter the following command in the Python console. The output should look like this:

    ```
    (pygeoan_cb)mdiener@mdiener-VirtualBox:~/venv$ python
    Python 2.7.3 (default,  Feb 27 2014, 19:58:35)
    [GCC 4.6.3] on linux2
    Type "help",  "copyright", "credits", or  "license" for more information.
    >> import numpy
    >> import pyproj
    ```

No errors, I hope. You have now successfully installed NumPy and pyproj.

 All sorts of errors could show up, so please take a look at the respective installation links to help you solve them:

For pyproj: https://pypi.python.org/pypi/pyproj/

For NumPy: http://www.numpy.org

How it works...

This easy installation works using the standard pip installation method. No tricks or special commands are needed. You need to simply execute the `pip install <library_name>` command and you are off to the races.

 Library names can be found by visiting the `https://pypi.python.org/pypi` web page if you are unsure of the exact name you want to install.

Installing shapely, matplotlib, and descartes

A large part of geospatial analysis and visualization is made possible using Shapely, matplotlib, GDAL, OGR, and descartes, which are installed later. Most of the recipes here will use these libraries extensively so setting them up is necessary to complete our exercises.

Shapely (`http://toblerity.org/shapely`) provides pure spatial analysis of geometries using the Cartesian coordinate system as is used by AutoCAD, for those of you familiar with CAD-like programs. The benefit of using a flat coordinate system is that all the rules of Euclidean geometry and analytic geometry are applied. For a quick refresher in the coordinate systems that we all learned in school, here is a little image to quickly jolt your memory.

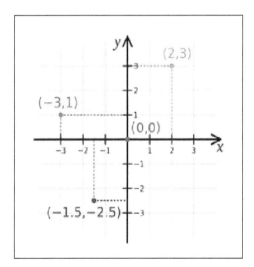

 Description: A Cartesian coordinate system demonstrating a flat plane to plot and measure geometry.

Illustration 1: Source: `http://en.wikipedia.org/wiki/Cartesian_coordinate_system`.

The classic overlay analysis and other geometric computations is where Shapely shines using the GEOS library as its workhorse in the background.

As for **matplotlib** (`http://matplotlib.org/`), it is the plotting engine that renders nice graphs and data to your screen as an image or **scalable vector graphic** (**svg**). The uses of matplotlib are only limited to your imagination. So, like the name partially implies, matplotlib enables you to plot your data on a graph or even on a map. For those of you familiar with MATLAB, you will find matplotlib quite similar in functionality.

The **descartes** library provides a nicer integration of Shapely geometry objects with Matplotlib. Here, you will see that descartes opens the `fill` and `patch` of matplotlib plots to work with the geometries from Shapely and saves you from typing them individually.

Getting ready

To prepare for installation, it is necessary to install some global packages, such as `libgeos_c`, as these are required by Shapely. NumPy is also a requirement that we have already met and is also used by Shapely.

Install the requirements of matplotlib from the command line like this:

```
$ sudo apt-get install freetype* libpng-dev libjpeg8-dev
```

These are the dependencies of matplotlib, which can be seen on a Ubuntu 14.04 machine.

How to do it...

Follow these instructions:

1. Run pip to install shapely:

   ```
   $ pip install shapely
   ```

2. Run pip to install matplotlib:

   ```
   $ pip install matplotlib
   ```

3. Finally, run pip to install descartes:

   ```
   $ pip install descartes
   ```

Another test to see if all has gone well is to simply enter the Python console and try to import the packages, and if no errors occur, your console should show an empty Python cursor. The output should look like what is shown in the following screenshot:

```
(pygeoan_cb)mdiener@mdiener-VirtualBox:~/venv$ python
Python 2.7.3 (default,  Feb 27 2014, 19:58:35)
[GCC 4.6.3] on linux2
Type "help",  "copyright", "credits", or  "license" for more information.
>>> import shapely
>>> import matplotlib
>>> import descartes
>>>

# type exit() to return
>>> exit()
```

If any errors occur, Python usually provides some good clues as to where the problem is located and there is always **Stack Overflow**. For example, have a look at http://stackoverflow.com/questions/19742406/could-not-find-library-geos-c-or-load-any-of-its-variants/23057508#2305750823057508.

How it works...

Here, the order in which you install the packages is very important. The descartes package depends on matplotlib, and matplotlib depends on NumPy plus freetype and libpng. This narrows you down to installing NumPy first, then matplotlib and its dependencies, and finally, descartes.

The installation itself is simple with pip and should be quick and painless. The tricky parts occur if `libgeos_c` is not installed properly, and you might need to install the `libgeos-dev` library.

Installing pyshp, geojson, and pandas

These specific libraries are for specific formats that make our life easier and simpler than using GDAL for some projects. pyshp will work with shapefiles, geojson with GeoJSON, and pandas with all other textual data types in a structured manner.

pyshp is pure Python and is used to import and export shapefiles; you can find the source code for pyshp here at `https://github.com/GeospatialPython/pyshp`. The pyshp library's sole purpose is to work with shapefiles. GDAL will be used to do most of our data's in/out needs, but sometimes, a pure Python library is simpler when working with shapefiles.

geojson is the name of a Python library and also a format, making it a little confusing to understand. The GeoJSON format (`http://geojson.org`) is becoming ever more popular and to this extent, we use the Python geojson library to handle its creation. You will find it on **Python Package Index** (**PyPI**) if you search for geojson. As you would expect, this will help us create all the different geometry types supported in the GeoJSON specifications.

pandas (`http://pandas.pydata.org`) is a data analysis library that structures your data in a spreadsheet-like manner for further computations. Since our geospatial data comes from a broad set of sources and formats, such as CSV, pandas helps work with the data with minimal effort.

Getting ready

Enter your virtual environment using the following command:

```
$ workon pygeoan_cb
```

Your prompt should now look like this:

```
(pygeoan_cb)mdiener@mdiener-VirtualBox:~$
```

How to do it...

The three installations are as follows:

1. Pyshp will first be installed by simply using pip as follows:

   ```
   $ pip install pyshp
   ```

2. Next, the geojson library will be installed using pip:

   ```
   $ pip install geojson
   ```

3. Finally, pip will install pandas:

   ```
   $ pip install pandas
   ```

To test your installation of pyshp, use the `import shapefile` type. The output should look like what is shown in the following output:

```
(pygeoan_cb)mdiener@mdiener-VirtualBox:~/venv$ python
Python 2.7.3 (default,  Feb 27 2014, 19:58:35)
[GCC 4.6.3] on linux2
Type "help",  "copyright", "credits", or  "license" for more information.
>> import shapefile
>> import geojson
>> import pandas
```

> The `import shapefile` statement imports the `pyshp` library; unlike the other libraries, the import name is not the same as the installation name.

How it works...

As seen in the other modules, we've used the standard installation pip package to execute installations. There are no other dependencies to worry about, making for fast progress.

Installing SciPy, PySAL, and IPython

SciPy is a collection of Python libraries, including SciPy library, matplotlib, pandas, SymPy, and IPython. The SciPy library itself is used for many operations, but we are particularly interested in the **spatial** module. This module can do many things including running a nearest neighbor query.

PySAL is a geospatial computing library that's used for spatial analysis. Creating models and running simulations directly from Python code are some of the many library functions that PySAL offers. PySAL is a library that, when put together with our visualization tools such as matplotlib, gives us a great tool.

IPython is a Python interpreter for a console that replaces the normal Python console you may be used to when running and testing Python code from your terminal. This is really just an advanced interpreter with some cool features, such as *Tab* completion, which means that beginners can get commands quickly by typing a letter and hitting *Tab*. The IPython notebooks help share code in the form of a web page, including code, images, and more without any installation.

Getting ready

The dependency jungle we looked at earlier is back and we need three more universal installations to our Ubuntu system using `apt-get install` as follows:

```
$ sudo apt-get install libblas-dev liblapack-dev gfortran
```

Windows and Mac users can use a full installer (`http://www.scipy.org/install.html`), such as Anaconda or Enthought Canopy, which will perform all the installation dependencies for you in one go.

Three dependencies are used for the SciPy installation. PySAL depends on SciPy so make sure to install SciPy first. Only IPython does not need any extra installations.

Start up your Python virtual environment with the following code:

```
mdiener@mdiener-VirtualBox:~$ workon pygeoan_cb
(pygeoan_cb)mdiener@mdiener-VirtualBox:~$
```

How to do it...

Let's look at these steps:

1. First, we'll install SciPy since PySAL depends on it. This will take a while to install; it took my machine 5 minutes to go through so take a break:

   ```
   $ pip install scipy
   ```

2. PySAL can be installed super quickly using pip:

   ```
   $ pip install pysal
   ```

3. As usual, we'd like to see whether everything's working, so let's fire up the Python shell as follows:

   ```
   (pygeoan_cb)mdiener@mdiener-VirtualBox:~$python
   >>> import scipy
   >>> import pysal
   >>>
   ```

4. IPython is to be installed globally or inside the virtual environment using pip as follows:

   ```
   $ pip install ipython
   ```

How it works...

SciPy and PySAL libraries are both geared to help accomplish various spatial analysis duties. The choice of tool is based on the task at hand, so make sure that you check which library offers what function at the command prompt as follows:

```
>>> from scipy import spatial
>>> help(spatial)
```

The output should look like what is shown in the following screenshot:

```
mdiener@mdiener-VirtualBox: ~/.venvs
Help on package scipy.spatial in scipy:

NAME
    scipy.spatial

FILE
    /home/mdiener/.venvs/pygeo_analysis_cookbook/local/lib/python2.7/site-packages/scipy/spatial/__init__.py

DESCRIPTION
    ===================================================================
    Spatial algorithms and data structures (:mod:`scipy.spatial`)
    ===================================================================

    .. currentmodule:: scipy.spatial

    Nearest-neighbor Queries
    =========================
    .. autosummary::
       :toctree: generated/

       KDTree      -- class for efficient nearest-neighbor queries
       cKDTree     -- class for efficient nearest-neighbor queries (faster impl.)
       distance    -- module containing many different distance measures

    Delaunay Triangulation, Convex Hulls and Voronoi Diagrams
    ===========================================================

    .. autosummary::
       :toctree: generated/

       Delaunay    -- compute Delaunay triangulation of input points
       ConvexHull  -- compute a convex hull for input points
       Voronoi     -- compute a Voronoi diagram hull from input points
:
```

Installing GDAL and OGR

Converting formats is boring, repetitive, and is one of the many, many responsibilities that the GDAL library provides, not to mention format transformations. However, GDAL also shines with regard to other geospatial functions, such as getting the current projections of a Shapefile or generating contours from elevation data. So, to only say that GDAL is a transformation library would be wrong; it really is so much more. The father of GDAL, Frank Warmerdam, deserves credit for starting it all off, and the GDAL project is now part of the **OSGEO** (**Open Source Geospatial Foundation**, refer to www.osgeo.org).

 The GDAL installation includes OGR; there is no extra installation required.

Currently, GDAL covers working with raster data, and OGR covers working with vector data. With GDAL 2.x now here, the two sides, raster and vector, are merged under one hat. GDAL and OGR are the so-called Swiss Army knives of geospatial data transformations, covering over 200 different spatial data formats.

Getting ready

GDAL isn't known to be the friendliest beast to install on Windows, Linux, or OSX. There are many dependencies and even more ways to install them. The descriptions are not all very straightforward. Keep in mind that this description is just one way of doing things and will not always work on all machines, so please refer to the online instructions for the latest and best ways to get your system up and running.

To start with, we will install some dependencies globally on our machine. After the dependencies have been installed, we will go into the global installation of GDAL for Python in our global site packages.

How to do it...

To globally install GDAL into our Python site packages, we will proceed with the following steps:

1. The following command is used when installing build and XML tools:

   ```
   $ sudo apt-get install -y build-essentiallibxml2-dev libxslt1-dev
   ```

2. Install the GDAL development files using the following command:

   ```
   $ sudo apt-get install libgdal-dev # install is 125MB
   ```

3. This following command will install GDAL package in the main Python package. This means that GDAL will be installed globally. The global installation of GDAL is usually not a bad thing since, as far as I am aware, there are no backward incompatible versions, which is very rare these days. The installation of GDAL directly and only in virtualenv is painful, to say the least, and if you are interested in attempting it, I've mentioned some links for you to try out.

   ```
   $ sudo apt-get install python-gdal
   ```

> If you would like to attempt the installation inside your virtual environment, please take a look at this Stack Overflow question at http://gis.stackexchange.com/questions/28966/python-gdal-package-missing-header-file-when-installing-via-pip.

4. To get GDAL in the Python virtual environment, we only need to run a simple virtualevnwrapper command:

   ```
   toggleglobalsitepackages
   ```

   Make sure you have your virtual environment activated as follows:

   ```
   mdiener@mdiener-VirtualBox:~$ workon pygeoan_cb
   (pygeoan_cb)mdiener@mdiener-VirtualBox:~$
   ```

5. Now, activate the global Python site packages in your current virtual environment:

```
(pygeoan_cb)mdiener@mdiener-VirtualBox:~$ toggleglobalsitepackages
enable global site-packages
```

6. The final check is to see if GDAL is available as follows:

```
$ python
>>> import gdal
>>>
```

7. No errors have been found and GDAL is ready for action.

Windows 7 plus users should use the OSGeo4W windows installer (`https://trac.osgeo.org/osgeo4w/`).Find the following section on the web page and download your Windows version in 32-bit or 64-bit. Follow the graphical installer instructions and the GDAL installation will then be complete.

 Windows users can also directly get binaries if all fails at `http://www.gisinternals.com/sdk/`. This installer should help avoid any other Windows specific problems that can arise and this site can help get you going in the right direction.

How it works...

The GDAL installation encompasses both the raster (GDAL) and vector (OGR) tools in one. Within the GDAL install are five modules that can be separately imported into your project depending on your needs:

```
>>> from osgeo import gdal
>>> from osgeo import ogr
>>> from osgeo import osr
>>> from osgeo import gdal_array
>>> from osgeo import gdalconst
>>> python
>>> import osgeo
>>> help(osgeo)
```

To see what packages are included with your Python GDAL installation, we use the Python built-in help function to list what the OSGeo module has to offer. This is what you should see:

```
NAME
    osgeo - # __init__ for osgeo package.
FILE
    /usr/lib/python2.7/dist-packages/osgeo/__init__.py
MODULE DOCS
    http://docs.python.org/library/osgeo
PACKAGE CONTENTS
    _gdal
    _gdal_array
    _gdalconst
    _ogr
    _osr
    gdal
    gdal_array
    gdalconst
    gdalnumeric
    ogr
    osr
DATA
    __version__ = '1.10.0'
    version_info = sys.version_info(major=2, minor=7, micro=3,
releaseleve...
VERSION
    1.10.0
(END)
```

At the time of writing this, the GDAL version is now bumped up to 2.0, and in developer land, this is old even before it gets printed. Beware that the GDAL 2.0 has compatibility issues and for this book, version 1.x.x is recommended.

See also

The http://www.gdal.org homepage is always the best place for reference regarding any information about it. The OSGEO includes GDAL as a supported project, and you can find more information on it at http://www.osgeo.org.

Installing GeoDjango and PostgreSQL with PostGIS

This is our final installation recipe and if you have followed along so far, you are ready for a simple, straightforward start to Django. Django is a web framework for professionals with deadlines, according to the Django homepage. The spatial part of it can be found in **GeoDjango**. GeoDjango is a contrib module installed with every Django installation therefore, you only need to install Django to get GeoDjango running. Of course, "geo" has its dependencies that were met in the previous sections. For reference purposes, take a look at this great documentation on the Django homepage at

```
https://docs.djangoproject.com/en/dev/ref/contrib/gis/install/#ref-
gis-install.
```

We will use PostgreSQL and PostGIS since they are the open source industry go-to spatial databases. The installations are not 100% necessary, but without them there is no real point because you then limit your operations, and they're definitely needed if you plan to store your spatial data in a spatial database. The combination of PostgreSQL and PostGIS is the most common spatial database setup for GeoDjango. This installation is definitely more involved and can lead to some hook-ups depending on your system.

Getting ready

To use GeoDjango, we will need to have a spatial database installed, and in our case, we will be using PostgreSQL with the PostGIS extension. GeoDjango also supports Oracle, Spatialite, and MySQL. The dependencies of PostGIS include GDAL, GEOS, PROJ.4, LibXML2, and JSON-C.

Start up your Python virtual environment as follows:

```
mdiener@mdiener-VirtualBox:~$ workon pygeoan_cb

(pygeoan_cb)mdiener@mdiener-VirtualBox:~$
```

Downloading the example code

You can download the example code files for all Packt books you have purchased from your account at http://www.packtpub.com. If you purchased this book elsewhere, you can visit http://www.packtpub.com/support and register to have the files e-mailed directly to you.

How to do it...

Follow these steps. These are taken from the PostgreSQL homepage for Ubuntu Linux:

1. Create a new file called `pgdg.list` using the standard gedit text editor. This stores the command to fire up your Ubuntu installer package:

    ```
    $ sudo gedit /etc/apt/sources.list.d/pgdg.list
    ```

2. Add this line to the file, save, and then close it:

    ```
    $ deb http://apt.postgresql.org/pub/repos/apt/ precise-pgdg main
    ```

3. Now, run the `wget` command for add the key:

    ```
    $ wget --quiet -O - https://www.postgresql.org/media/keys/
    ACCC4CF8.asc | \ sudo apt-key add -
    ```

4. Run the `update` command to actualize your installer packages:

    ```
    $ sudo apt-get update
    ```

5. Run the `install` command to actually install PostgreSQL 9.3:

    ```
    $ sudo apt-get install postgresql-9.3
    ```

6. To install PostGIS 2.1, we will have one unmet dependency, `libgdal1`, so go ahead and install it:

    ```
    $ sudo apt-get install libgdal1
    ```

7. Now we can install PostGIS 2.1 for PostgreSQL 9.3 on our machine:

    ```
    $ sudo apt-get install postgresql-9.3-postgis-2.1
    ```

8. Install the PostgreSQL header files:

    ```
    $ sudo apt-get install libpq-dev
    ```

9. Finally, install the `contrib` module with contributions:

    ```
    $ sudo apt-get install postgresql-contrib
    ```

10. Install the Python database adapter, `psycopg2`, to connect to your PostgreSQL database from Python:

    ```
    $ sudo apt-get install python-psycopg2
    ```

11. Now we can create a standard PostgreSQL database as follows:

    ```
    (pygeoan_cb)mdiener@mdiener-VirtualBox:~$ createdb
    [NewDatabaseName]
    ```

12. Using the `psql` command-line tool, we can create a PostGIS extension to our newly created database to give it all the PostGIS functions as follows:

```
(pygeoan_cb)mdiener@mdiener-VirtualBox:~$ psql -d
[NewDatabaseName] -c "CREATE EXTENSION postgis;"
```

13. Moving on, we can finally install Django in one line directly in our activated virtual environment:

```
$ pip install django
```

14. Test out your install of Django and GDAL and, as always, try to import them as follows:

```
>>> from django.contrib.gis import gdal

>>> gdal.HAS_GDAL

True
```

Windows users should be directed to the PostgreSQL Windows (`http://www.postgresql.org/download/windows/`) binaries provided by EnterpriseDB (`http://www.enterprisedb.com/products-services-training/pgdownload#windows`). Download the correct version and follow the installer instructions. PostGIS is also included in the list of extensions that you can directly install using the installer.

How it works...

Installations using the apt-get Ubuntu installer and the Windows installers are simple enough in order to have PostgreSQL, PostGIS, and Django up and running. However, the inner workings of the installers are beyond the scope of this book.

There's more...

To summarize all the installed libraries, take a look at this table:

Library name	Description	Reason to install
NumPy	This adds support for large multidimensional arrays and matrices	It is a requirement for many other libraries
pyproj	This handles projections	It transforms projections
shapely	This handles geospatial operations	It performs fast geometry manipulations and operations
matplotlib	This plots libraries	It provides a quick visualization of results
descartes	This uses Shapely or GeoJSON objects as matplotlib paths and patches	It speedily plots geo-data
pandas	This provides high-performance data structures and data analysis	It performs data manipulation, CSV creation, and data manipulation

Library name	Description	Reason to install
SciPy	This provides a collection of Python libraries for scientific computing	It has the best collection of necessary tools
PySAL	This contains a geospatial analysis library	It performs a plethora of spatial operations (optional)
IPython	This provides interactive Python computing	It is a helpful notebook to store and save your scripts (optional)
Django	This contains a web application framework	It is used for our demo web application in *Chapter 11, Web Analysis with GeoDjango*
pyshp	This provides pure Python shapefile manipulation and generation	It helps input and output shapefiles
GeoJSON	This contains the JSON format for spatial data	It facilitates the exchange and publication of this format
PostgreSQL	This is a relational database	It helps store spatial data
PostGIS	This is the spatial extension to PostgreSQL	It stores and performs spatial operations on geographic data in PostgreSQL

2
Working with Projections

In this chapter, we will cover the following topics:

- ▶ Discovering projection(s) of a Shapefile or GeoJSON dataset
- ▶ Listing projection(s) from a WMS server
- ▶ Creating a projection definition for a Shapefile if it does not exist
- ▶ Batch setting the projection definition of a folder full of Shapefiles
- ▶ Reprojecting a Shapefile from one projection to another

Introduction

Working with projections, in my opinion, is not too exciting but they're very important, and your ability to deal with them in any application is crucial.

The goal of this chapter is to provide some common predata screening or transformation steps to get your data in shape or, better yet, in position for geospatial analysis. We cannot always perform analysis on multiple datasets that are in different coordinate systems without the risk of achieving inconsistent results, such as data positional inaccuracies. Therefore, it is a best practice to work on data in the same coordinate system, such as EPSG:4326, when working on a global scale, or use a local coordinate system for your region that will provide you the most accurate results.

European Petroleum Survey Group or **EPSG** codes have decided to give all coordinate systems a number code to simplify finding and sharing projection information. Coordinate systems are described by their definitions, which are stored in text files of various formats. These text files are designed to be computer-readable formats, specifically for individual GIS desktop software packages, such as QGIS or ESRI ArcMap or for your web/scripting applications.

The EPSG code 4326 represents **World Geographic System 1984 (WGS 84)** and is a **geographic coordinate system** with longitude and latitude (*x, y*) units (refer to the following image). The geographic coordinate system represents the Earth as a sphere, as in this image, and the unit of measurement is degrees.

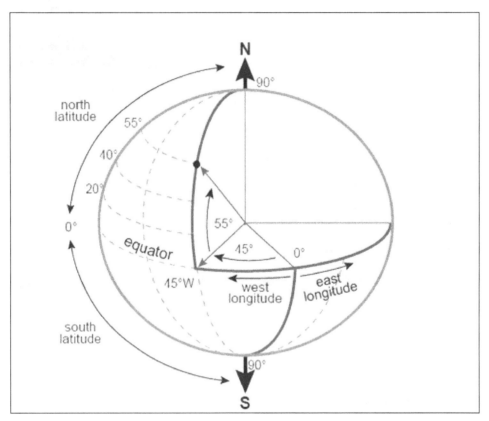

Illustration 1: Geographic Coordinate System
(http://kartoweb.itc.nl/geometrics/coordinate%20systems/coordsys.html)

The second type of coordinate system is a **projected coordinate system**, which is a two-dimensional flat plane with constant areas, lengths, or angles that are measured on an *x* and *y* grid. **EPSG:3857 Pseudo-Mercator** is such a projected coordinate system where the units are in meters with correct lengths but the angles and areas are distorted. In any given projected coordinate system, only two of the three properties, *area*, *distance*, or *angles*, can be correctly represented on a single map. The **Universal Transverse Mercator** (**UTM**) coordinate reference system divides the world into 60 zones (refer to the following image):

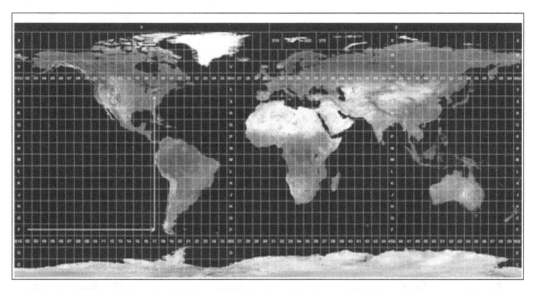

Illustration 2: Projected Coordinate System UTM (http://en.wikipedia.org/wiki/Universal_Transverse_Mercator_coordinate_system#mediaviewer/File:Utm-zones.jpg)

 Note that you have to enter your Python virtual environment using the `workon pygeoan_cb` command.

Discovering projection(s) of a Shapefile or GeoJSON dataset

Remember that all data is stored in a coordinate system, no matter what the data source is. It is your job to figure this out using a simple approach outlined in this section. We will take a look at two different data storage types: a **Shapefile** and a **GeoJSON** file. These two formats contain geometries, such as points, lines, or polygons, and their associated attributes. For example, a tree would be stored as a point geometry with attributes, such as height, age, and species, Each of these data types store their projection data differently and, therefore, require different methods to discover their projection information.

Now a quick introduction to what a Shapefile is: a Shapefile is not a single file but a minimum of three files, such as `.shp`, `.shx`, and, `.dbf`, all of which have the same name. For example, `world_borders.shp`, `world_borders.shx` and `world_borders.dbf` make up one file. The `.shp` file stores geometry, `.dbf` stores a table of attribute values, and `.shx` is the index table that connects geometry to an attribute value as a lookup table.

A Shapefile should come with a very important fourth text file called `world_borders.prj`. The **.prj** stands for **projection information** and contains the projection definition of the Shapefile in a plain text format. As crazy as it sounds, you can still find and download tons of data being delivered today without this `.prj` file. You can do this simply by opening this `.prj` file in a text editor, such as *Sublime Text* or *Notepad++*, where you can read about the projection definition in order to determine the files coordinate system.

The `.prj` file is a plain text file that can easily be generated for the wrong coordinate system if you are not careful. The wrong projection definition can cause problems with your analysis and transformations. We will see how to correctly assess a Shapefile's projection information.

GeoJSON is a single file stored in plain text. The GeoJSON standard (`http://www.geojson.org`) is based on the JSON standard. The coordinate reference information is, in my experience, often *not* included and the default is **WGS 84** and **EPSG:4326**, where the coordinates are stored in the x, y, z format and in this *exact order*.

Mixing *x* and *y* for *y*, *x* can happen and when it does, your data will most likely end up in the ocean, so always remember that order matters:

x = longitude

y = latitude

z = height

The following is how the GeoJSON CRS information looks if it is presented in `FeatureCollection`:

```
{
"type": "FeatureCollection",
"crs": { "type": "name", "properties": { "name":
  "urn:ogc:def:crs:OGC:1.3:CRS84" } },
...
```

Getting ready

The first thing is to head over to `https://github.com/mdiener21/python-geospatial-analysis-cookbook` and download the entire source code and geodata in one go. The download icon is located on the bottom right-hand side and is labeled as **Download ZIP**. If you are a GitHub user, you can, of course, clone the repository. Beware that it is a download that's a little over 150 MB. Inside the repository, you will find each chapter with the following three folders: `/geodata/` to store data, `/code/` to store code scripts that have been completed, and an empty folder called `/working/` for you to create your self-written code scripts. The structure looks like this:

```
/ch01/
-------/code
-------/geodata
-------/working
/ch02/
-------/code
-------/geodata
-------/working
...
```

The source of the data used in this recipe comes from the City of Vancouver, BC, Canada, which is located at `http://data.vancouver.ca/datacatalogue/index.htm` (Vancouver schools).

When downloading data from an Internet source, always look around for metadata descriptions about the projection information so that you know a little about your data's history and its source before you begin working with it. Most data today is publicly available in EPSG:4326 WGS 84 or EPSG:3857 Web Pseudo-Mercator. Data stemming from a government resource is most likely stored in local coordinate systems.

How to do it...

We are going to start with our Shapefile and identify the coordinate system it is stored in using the GDAL library that imports the OGR module:

Note that we have assumed that your Shapefile has a `.prj` file. If not, this process will not work.

1. Create a new Python file named as `ch02_01_show_shp_srs.py` in your `/ch02/working/` directory, and add the following code:

```python
#!/usr/bin/env python
# -*- coding: utf-8 -*-

from osgeo import ogr
shp_driver = ogr.GetDriverByName('ESRI Shapefile')
shp_dataset = shp_driver.Open(r'../geodata/schools.shp')
shp_layer = shp_dataset.GetLayer()
shp_srs = shp_layer.GetSpatialRef()
print shp_srs
```

2. Now save the file and run the `ch02_01_show_shp_srs.py` script from the command line:

```
$ python ch02-01-show_shp_srs.py
PROJCS["NAD_1983_UTM_Zone_10N",
    GEOGCS["GCS_North_American_1983",
        DATUM["North_American_Datum_1983",
            SPHEROID["GRS_1980",6378137,298.257222101]],
        PRIMEM["Greenwich",0],
        UNIT["Degree",0.017453292519943295]],
    PROJECTION["Transverse_Mercator"],
    PARAMETER["latitude_of_origin",0],
    PARAMETER["central_meridian",-123],
    PARAMETER["scale_factor",0.9996],
    PARAMETER["false_easting",500000],
    PARAMETER["false_northing",0],
    UNIT["Meter",1]]
```

You should see the preceding text print out on your screen, showing information on the `.prj` projection.

 Note that we could also simply open the `.prj` file with a text editor and view this information as well.

Now, we will take a look at a GeoJSON file to see the projection information if it's available.

3. Determining the coordinate system of a **GeoJSON** file is a little harder since we must make one of two assumptions where the first case is the standard case and most common:

    1. No CRS is explicitly defined inside the GeoJSON, so we assume EPSG:4326 WGS 84 is the coordinate system.

    2. CRS is defined explicitly and is correct.

4. Create a new Python file named `ch02_02_show_geojson_srs.py` in your `/ch02/working/` directory, and add the following code:

```
#!/usr/bin/env python
# -*- coding: utf-8 -*-

import json

geojson_yes_crs = '../geodata/schools.geojson'
geojson_no_crs = '../geodata/golfcourses_bc.geojson'

with open(geojson_no_crs) as my_geojson:
    data = json.load(my_geojson)

# check if crs is in the data python dictionary data
# if yes print the crs to screen
# else print NO to screen and print geojson data type
if 'crs' in data:
    print "the crs is : " + data['crs']['properties']['name']
else:
    print "++++++ no crs tag in file+++++"
    print "++++++ assume EPSG:4326 ++++++"
    if "type" in data:
        print "current GeoJSON data type is :" +
data['type']
```

5. The script is set up to run on the `geojson_no_crs` variable set in the GeoJSON `golfcourses_bc.geojson` file. The source of this data is **OpenStreetMap**, which is exported using the **Overpass API** that's located at `http://overpass-turbo.eu/`. Now, run the `ch02_02_show_geojson_srs.py` script and you should see this output for our first file:

```
$ python ch02_02_show_geojson_crs.py

++++++ no crs tag in file+++++

        ++++++ assume EPSG:4326 ++++++

current GeoJSON data type is :FeatureCollection
```

> If no CRS is inside our GeoJSON file, we'll assume it has a projection of EPSG:4326. To check this, you will need to look at the coordinates listed inside the file and see if they fall within bounds, such as `-180.0000`, `-90.0000`, `180.0000`, and `90.0000`. If so, we'll assume that the dataset is truly EPSG:4326 and open the data in QGIS to check this.

6. Now, go into the code and edit line 10, change the variable from `geojson_no_crs` to `geojson_yes_crs`, and rerun the `ch02_02_show_geojson_srs.py` code file:

```
$ python ch02_02_show_geojson_crs.py
the crs is : urn:ogc:def:crs:EPSG::26910
```

You should now see the preceding output printed on screen.

How it works...

Beginning with the Shapefile, we've used the OGR library to help us quickly discover the EPSG code information of our Shapefile as follows:

1. Begin with importing the OGR module as follows:

```
from osgeo import ogr
```

2. Activate the OGR Shapefile driver:

```
shp_driver = ogr.GetDriverByName('ESRI Shapefile')
```

3. Open the Shapefile with OGR:

```
shp_dataset = shp_driver.Open(r'../geodata/schools.shp')
```

4. Access the layer information with `GetLayer()`:

```
shp_layer = shp_dataset.GetLayer()
```

5. Now we can get the coordinate information using the `GetSpatialRef()` function:

```
shp_srs = shp_layer.GetSpatialRef()
```

6. Finally, print the spatial reference system on screen:

```
print shp_srs
```

The GeoJSON file was a little harder to tackle when we used the Python JSON module to look for the `crs` key and print out its value on the screen, if it existed.

We could have simply replaced the first example code with the GeoJSON driver and we would get the same result. However, not all GeoJSON files include projection information. The OGR driver will always output WGS 84 as the coordinate system by default which, in our `no_geojson_crs.geojson` example file, is wrong. This can lead to confusion for new users. The important thing to note is to check your data, have a look at the coordinate values, and see if they fit in a defined coordinate range of values. To explore codes, or if you enter a code that you have and want to see the area it covers on a live web map, refer to `http://epsg.io`.

First, we'll import the standard Python JSON module, and then set two variables to store both our GeoJSON files. Next, we'll open one file, the `golfcourses_bc.geojson` file, and load the GeoJSON file into a Python object. Then, all we need to do is check to see whether the `crs` key is in the GeoJSON; if it is, we'll print out its value. If not, we'll simply print to screen that `crs` is not available and the GeoJSON data type.

The GeoJSON default CRS is WGS 84 EPSG:4326, which means that we are dealing with latitude and longitude values. The values must fall within the bounds of `-180.0000`, `-90.0000`, `180.0000`, and `90.0000` to qualify.

There's more...

Here are some projection definition examples for your reference:

1. The code for the ESRI Well-Known Text (stored with Shapefile as `ShapefileName.prj`) is as follows:

   ```
   GEOGCS["GCS_WGS_1984",DATUM["D_WGS_1984",SPHEROID["WGS_198
   4",6378137,298.257223563]],PRIMEM["Greenwich",0],UNIT["Degr
   ee",0.017453292519943295]]
   ```

2. The code for the OGC Well-Known Text of the same coordinate system as EPSG:4326 is as follows:

   ```
   GEOGCS["WGS 84",DATUM["WGS_1984",SPHEROID["WGS
   84",6378137,298.257223563,AUTHORITY["EPSG","7030"]],AUTHORITY["EPS
   G","6326"]],PRIMEM["Greenwich",0,AUTHORITY["EPSG","8901"]],UNIT["d
   egree",0.01745329251994328,AUTHORITY["EPSG","9122"]],AUTHORITY["EP
   SG","4326"]]
   ```

3. The code for the Proj4 format, which also shows `EPSG:4326`, is as follows:

   ```
   +proj=longlat +ellps=WGS84 +datum=WGS84 +no_defs
   ```

See also

The web page at `http://www.spatialreference.org` is a great place to get coordinates for any projection you desire by simply selecting the destination coordinate system that you like, zooming in on the map, and then copying and pasting the coordinates. Later on, we will use the `http://spatialreference.org/` API to get the EPSG definition to create our own `.prj` file for a Shapefile.

Listing projection(s) from a WMS server

The **Web Mapping Service** (**WMS**), which can be found at `https://en.wikipedia.org/wiki/Web_Map_Service`, is fun since most service providers provide data in several coordinate systems and you can then specify which one you would like. However, you can't reproject or transform the WMS into some other system that the service provider does not provide, which means that you can only use the coordinate system that is provided. The following is an example of a WMS `getCapabilities` request (`http://gis.ktn.gv.at/arcgis/services/INSPIRE/INSPIRE/MapServer/WmsServer?service=wms&version=1.3.0&request=getcapabilities`), showing a list of the five available coordinate systems from a WMS service:

Getting ready

The WMS service URL that we will use is `http://ogc.bgs.ac.uk/cgi-bin/BGS_1GE_Geology/wms?service=WMS&version=1.3.0&request=GetCapabilities`. This is from the British Geological Survey, titled *OneGeology Europe geology*.

 For a list of WMS servers that are available worldwide, refer to Skylab Mobile Systems at `http://www.skylab-mobilesystems.com/en/wms_serverlist.html`. Also, take a look at `http://geopole.org/`.

We will use a library called OWSLib. This library is a great package for working with OGC web services such as WMS, as follows:

```
Pip install owslib
```

How to do it...

Let's go through these steps to retrieve the projections that a WMS server provides and print the available EPSG codes to screen:

1. Create a new Python file named `ch02_03_show_wms_srs.py` in your `/ch02/code/working/` directory, and add the following code:

```python
#!/usr/bin/env python
# -*- coding: utf-8 -*-

from owslib.wms import WebMapService

url = "http://ogc.bgs.ac.uk/cgi-bin/BGS_1GE_Geology/wms"

get_wms_url = WebMapService(url)

crs_list = get_wms_url.contents['GBR_Kilmarnock_BGS_50K_CompressibleGround'].crsOptions

print crs_list
```

2. Now, run the `ch02_03_show_wms_srs.py` script and you should see the following screen output:

```
$ python ch02_03_show_wms_srs.py
['EPSG:3857', 'EPSG:3034', 'EPSG:4326', 'EPSG:3031', 'EPSG:27700',
'EPSG:900913', 'EPSG:3413', 'CRS:84', 'EPSG:4258']
```

How it works...

Determining information on the WMS projection involves using the OWSLib library. This is quite a powerful way to get all kinds of OGC web service information from your client. The code simply takes in the WMS URL to retrieve the WMS information. The content of the response is called, and we are able to access the `crsOptions` attribute to list out all the available EPSG codes.

Creating a projection definition for a Shapefile if it does not exist

You recently downloaded a Shapefile from an Internet resource and saw that the `.prj` file was not included. You do know, however, that the data is stored in the EPSG:4326 coordinate system as stated on the website from where you downloaded the data. Now the following code will create a new `.prj` file.

Getting ready

Start up your Python virtual environment with the `workon pygeo_analysis_cookbook` command:

How to do it...

In the following steps, we will take you through creating a new `.prj` file to accompany our Shapefile. The `.prj` extension is necessary for many spatial operations performed by a desktop GIS, web service, or script:

1. Create a new Python file named `ch02_04_write_prj_file.py` in your `/ch02/code/working/` directory and add the following code:

```python
#!/usr/bin/env python
# -*- coding: utf-8 -*-

import urllib
import os

def get_epsg_code(epsg):
    """
    Get the ESRI formatted .prj definition
    usage get_epsg_code(4326)

    We use the http://spatialreference.org/ref/epsg/4326/esriwkt/
    """
```

```
    f=urllib.urlopen("http://spatialreference.org/ref/epsg/{0}/
esriwkt/".format(epsg))
    return (f.read())


# Shapefile filename must equal the new .prj filename
shp_filename = "../geodata/UTM_Zone_Boundaries"



# Here we write out a new .prj file with the same name
# as our Shapefile named "schools" in this example

with open("../geodata/{0}.prj".format(shp_filename), "w")
as prj:
    epsg_code = get_epsg_code(4326)
    prj.write(epsg_code)
    print "done writing projection definition to EPSG: " +
epsg_code
```

2. Now, run the ch02_04_write_prj_file.py script:

   $ python ch02_04_write_prj_file.py

3. You should see the following screen output:

   done writing projection definition UTM_Zone_Boundaries.prj to EPSG:4326

4. Inside your folder, you should see a new .prj file created with the same name as the Shapefile.

How it works...

We first wrote a function to fetch our projection definition text using the http://spatialreference.org/ API by passing the EPSG code value. The function returns a textual description of the EPSG code information using the esriwkt formatting style, indicating ESRI Well-Known Text, which is the format that the ESRI software uses to store the .prj file information.

Then we need to input the Shapefile name because the filename of .prj must be equal to the Shapefile name.

In the last step, we'll create the .prj file using shp_filename that is specified, and call the function that we wrote to get the text definition of the coordinate reference system.

Batch setting the projection definition of a folder full of Shapefiles

Working with one Shapefile is fine but working with tens or hundreds of files is something else. In such a scenario, we'll need automation to get a job done fast.

We have a folder that contains several Shapefiles that are all in the same coordinate system but do not have a .prj file. We want to create a .prj file for each Shapefile in the current directory.

This script is a modified version of the previous code example that could write a .prj file for a single Shapefile into a batch process that can run over several Shapefiles.

How to do it...

We have a folder with many Shapefiles and we would like to create a new .prj file for each Shapefile in this folder, so let's get started:

1. Create a new Python file named ch02_05_batch_shp_prj.py in your /ch02/ code/working/ directory and add the following code:

```python
#!/usr/bin/env python
# -*- coding: utf-8 -*-

import urllib
import os
from osgeo import osr


def create_epsg_wkt_esri(epsg):
    """
    Get the ESRI formatted .prj definition
    usage create_epsg_wkt(4326)

    We use the
http://spatialreference.org/ref/epsg/4326/esriwkt/

    """
    spatial_ref = osr.SpatialReference()
    spatial_ref.ImportFromEPSG(epsg)

    # transform projection format to ESRI .prj style
    spatial_ref.MorphToESRI()
```

```
    # export to WKT
    wkt_epsg = spatial_ref.ExportToWkt()

    return wkt_epsg


# Optional method to get EPGS as wkt from a web service
def get_epsg_code(epsg):
    """

    Get the ESRI formatted .prj definition
    usage get_epsg_code(4326)

    We use the http://spatialreference.org/ref/epsg/4326/esriwkt/

    """
    web_url = "http://spatialreference.org/ref/epsg/{0}/esriwkt/"
.format(epsg)
    f = urllib.urlopen(web_url)
    return f.read()


# Here we write out a new .prj file with the same name
# as our Shapefile named "schools" in this example
def write_prj_file(folder_name, shp_filename, epsg):
    """

    input the name of a Shapefile without the .shp
    input the EPSG code number as an integer

    usage  write_prj_file(<ShapefileName>,<EPSG CODE>)

    """

    in_shp_name = "/{0}.prj".format(shp_filename)
    full_path_name = folder_name + in_shp_name

    with open(full_path_name, "w") as prj:
        epsg_code = create_epsg_wkt_esri(epsg)
        prj.write(epsg_code)
        print ("done writing projection definition : " +
epsg_code)


def run_batch_define_prj(folder_location, epsg):
    """
```

```
                input path to the folder location containing
                all of your Shapefiles

                usage  run_batch_define_prj("../geodata/no_prj")

                """

                # variable to hold our list of shapefiles
                shapefile_list = []

                # loop through the directory and find shapefiles
                # for each found shapefile write it to a list
                # remove the .shp ending so we do not end up with
                # file names such as .shp.prj
                for shp_file in os.listdir(folder_location):
                    if shp_file.endswith('.shp'):
                        filename_no_ext = os.path.splitext(shp_file)[0]
                        shapefile_list.append(filename_no_ext)

                # loop through the list of shapefiles and write
                # the new .prj for each shapefile
                for shp in shapefile_list:
                    write_prj_file(folder_location, shp, epsg)


        # Windows users please use the full path
        # Linux users can also use full path
        run_batch_define_prj("c:/02_DEV/01_projects/04_packt/
        ch02/geodata/no_prj/", 4326)
```

How it works...

Using the standard `urllib` Python module, we can access an EPSG code via the Web and write this definition to a `.prj` file. We need to create a list of Shapefiles that we want to define `.prj` for and then create a `.prj` file for each Shapefile in this list.

The `get_epsg_code(epsg)` function returns the ESPG code text definition that we need. The `write_prj_file(shp_filename, epsg)` function takes two parameters, the Shapefile name and the EPSG code, writing out the `.prj` file to disk.

Next, we'll create an empty list to store the list of Shapefiles, switch to the directory where the Shapefiles are stored, and then list all the Shapefiles that currently in this directory.

Our `for` loop then populates the Shapefile list with the filenames without the `.shp` extension. Finally, the last `for` loop takes us through each Shapefile and calls our function to write each `.prj` file for each Shapefile in the list.

Reprojecting a Shapefile from one projection to another

Working with spatial data from multiple sources leads to data that's most likely from multiple regions on Earth with multiple coordinate systems. To perform consistent spatial analysis, we should transform all our input data into the same coordinate system. This means reprojecting your Shapefile into your chosen working coordinate system.

In this recipe, we will reproject a single Shapefile from ESPG:4326 into a web mercator system EPSG:3857 for use in a web application.

How to do it...

Our goal is to reproject a given Shapefile from one coordinate system to another; the steps to do this are as follows:

1. Create a new Python file named `ch02_06_re_project_shp.py` in your `/ch02/code/working/` directory and add the following code:

```python
#!/usr/bin/env python
# -*- coding: utf-8 -*-

import ogr
import osr
import os

shp_driver = ogr.GetDriverByName('ESRI Shapefile')

# input SpatialReference
input_srs = osr.SpatialReference()
input_srs.ImportFromEPSG(4326)

# output SpatialReference
output_srs = osr.SpatialReference()
output_srs.ImportFromEPSG(3857)

# create the CoordinateTransformation
coord_trans = osr.CoordinateTransformation(input_srs, output_srs)

# get the input layer
input_shp = shp_driver.Open(r'../geodata/UTM_Zone_Boundaries.shp')
in_shp_layer = input_shp.GetLayer()

# create the output layer
```

```
output_shp_file =
r'../geodata/UTM_Zone_Boundaries_3857.shp'
# check if output file exists if yes delete it
if os.path.exists(output_shp_file):
    shp_driver.DeleteDataSource(output_shp_file)

# create a new Shapefile object
output_shp_dataset =
shp_driver.CreateDataSource(output_shp_file)

# create a new layer in output Shapefile and define its geometry
type
output_shp_layer =
output_shp_dataset.CreateLayer("basemap_3857",
geom_type=ogr.wkbMultiPolygon)

# add fields to the new output Shapefile
# get list of attribute fields
# create new fields for output
in_layer_def = in_shp_layer.GetLayerDefn()
for i in range(0, in_layer_def.GetFieldCount()):
    field_def = in_layer_def.GetFieldDefn(i)
    output_shp_layer.CreateField(field_def)

# get the output layer's feature definition
output_layer_def = output_shp_layer.GetLayerDefn()

# loop through the input features
in_feature = in_shp_layer.GetNextFeature()
while in_feature:
    # get the input geometry
    geom = in_feature.GetGeometryRef()
    # reproject the geometry
    geom.Transform(coord_trans)
    # create a new feature
    output_feature = ogr.Feature(output_layer_def)
    # set the geometry and attribute
    output_feature.SetGeometry(geom)
    for i in range(0, output_layer_def.GetFieldCount()):
        output_feature.SetField(output_layer_def.GetFieldDefn(i)
.GetNameRef(), in_feature.GetField(i))
    # add the feature to the shapefile
    output_shp_layer.CreateFeature(output_feature)
    # destroy the features and get the next input feature
    output_feature.Destroy()
```

```
        in_feature.Destroy()
        in_feature = in_shp_layer.GetNextFeature()

    # close the shapefiles
    input_shp.Destroy()
    output_shp_dataset.Destroy()

    spatialRef = osr.SpatialReference()
    spatialRef.ImportFromEPSG(3857)

    spatialRef.MorphToESRI()
    prj_file = open('UTM_Zone_Boundaries.prj', 'w')
    prj_file.write(spatialRef.ExportToWkt())
    prj_file.close()
```

2. Now we can run our code from the command line as follows:

   ```
   $ python ch02_06_re_project_shp.py
   ```

3. We now have a new Shapefile called `UTM_Zone_Boundaries_3857.shp` that is in the `EPSG:3857` coordinate system and is ready for further use.

How it works...

The `osgeo`, `ogr`, and `osr` modules do the heavy lifting and the code required to reproject a Shapefile is quite verbose. It works by going through each geometry and transforming it individually into the new coordinate system.

Starting with the driver for the ESRI Shapefile, we'll work at setting our input and output **Spatial Reference System** (**SRS**) so that we can transform the two.

We need to copy each feature's geometry and its attributes from the old Shapefile into the new one as we transform each geometry. Finally, we'll close the input and output Shapefile with the `Destroy()` function.

See also

Using code is not always the best or fastest way to reproject a Shapefile. Another method that you could use is the `ogr2ogr` command-line tool that will simply reproject a Shapefile in one line. You could pipe this one-liner into a Python script and batch reproject many Shapefiles:

```
ogr2ogr -t_srs EPSG:4326 outputwith4236.shp input.shp
```

The GDAL library comes with several other very useful and helpful command-line functions that are worth checking out.

.

3

Moving Spatial Data from One Format to Another

In this chapter, we will cover the following topics:

- ▶ Converting a Shapefile to a PostGIS table using ogr2ogr
- ▶ Batch importing a folder of Shapefiles into PostGIS using ogr2ogr
- ▶ Batch exporting a list of tables from PostGIS to Shapefiles
- ▶ Converting an OpenStreetMap (OSM) XML to a Shapefile
- ▶ Converting a Shapefile (vector) to a GeoTiff (raster)
- ▶ Converting a raster (GeoTiff) to a vector (Shapefile) using GDAL
- ▶ Creating a Shapefile from point data stored in Microsoft Excel
- ▶ Converting an ESRI ASCII DEM to an image height map

Introduction

Geospatial data comes in hundreds of formats and massaging this data from one format to another is a simple task. The ability to convert between data types, such as rasters or vectors, belongs to data wrangling tasks and can be used for better geospatial analysis. Here is an example of a raster and vector dataset so that you can take a look at what I am talking about:

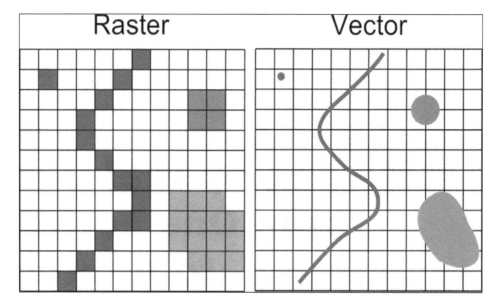

The best practice methodology is to run analysis functions or models over data stored in a common format, such as a PostgreSQL PostGIS database, or a set of Shapefiles in a common coordinate system. For example, running an analysis on input data stored in multiple formats is also possible, but you can expect to find the devil in the detail if something goes wrong or your results are not as you expected them to be.

This chapter looks at some common data formats and demonstrates how to move these from one format to another with the help of the most common tools.

Converting a Shapefile to a PostGIS table using ogr2ogr

The simplest way to transform data from one format to another is to directly use the *ogr2ogr* tool that comes with the installation of GDAL. This powerful tool can convert over 200 geospatial formats. In this solution, we will execute the *ogr2ogr* utility from within a Python script to perform generic vector data conversions. The Python code is, therefore, used to execute this command-line tool and pass around variables so that you can create your own scripts for data imports or exports.

Using this tool is also recommended if you are not really interested in coding too much and simply want to get the job done to move your data. A pure Python solution is, of course, possible, but it's definitely more orientated to the needs of developers (or a Python purist). Since this book is aimed at developers, analysts, or researchers, this kind of a recipe is simple and yet extensible.

Getting ready

To run this script, you will need the GDAL utilities application installed on your system. Windows users can visit OSGeo4W (`http://trac.osgeo.org/osgeo4w`) and download the 32-bit or 64-bit Windows installer. Simply double-click on the installer to start the script as follows:

1. Navigate to the bottom option, **Advanced Installation | Next**.

2. Click on **Next** to download the GDAL utilities from the Internet (first default option).

3. Click on **Next** to accept the default location of the path or change it to your liking.

4. Click on **Next** to accept the location of local saved downloads (default).

5. Click on **Next** to accept a direct connection (default).

6. Click on **Next** to select the default download site.

7. Now, you can finally see the menu. Click on **+** to open the **Commandline_Utilities** tab, and you should see what is shown in this screenshot:

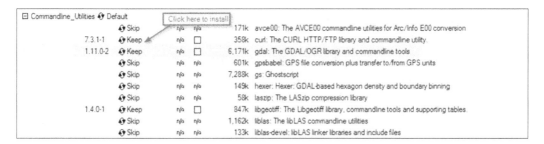

8. Now, select **gdal: The GDAL/OGR library and commandline tools** to install it.

9. Click on **Next** to start downloading and install.

Ubuntu/Linux users can use the following steps to install the GDAL utilities:

1. Execute this simple one-line command:

```
$ sudo apt-get install gdal-bin
```

This will get you up and running so that you can execute `ogr2ogr` directly from your terminal.

The Shapefile to be imported is located in your `/ch02/geodata/` folder if you've downloaded the entire source and code from GitHub at `https://github.com/mdiener21/python-geospatial-analysis-cookbook/`. The Vancouver open geodata portal (`http://data.vancouver.ca/datacatalogue/index.htm`) is our source that provides a dataset of local bikeways.

2. Next, let's set up our PostgreSQL database with the PostGIS extension. To do this, we'll first create a new user to manage our new database and tables as follows:

```
Sudo su createuser  -U postgres -P pluto
```

3. Enter a password for the new role.

4. Enter the password again for the new role.

5. Enter a password for the `postgres` user as you will create the user using this `postgres` user.

6. The `-P` option prompts you to give the new user, called `pluto`, a password. For the following examples, our password is `stars`; I would recommend a much more secure password for your production database.

 Windows users can navigate to the `c:\Program Files\ PostgreSQL\9.3\bin\` folder and execute the following command, and follow the on-screen instructions as you did earlier:

```
Createuser.exe -U postgres -P pluto
```

7. To create the database, we will use the same command-line `createdb` command as the `postgres` user to create a database named `py_geoan_cb`, and assign the `pluto` user to be the database owner. Here is the command to do this:

```
$ sudo su createdb -O pluto -U postgres py_geoan_cb
```

 Windows users can visit `c:\Program Files\PostgreSQL\9.3\ bin\` and execute the `createdb.exe` command as follows:

```
createdb.exe -O pluto -U postgres py_geoan_cb
```

Next, we'll create the PostGIS extension for our newly created database:

```
psql -U postgres -d py_geoan_cb -c "CREATE EXTENSION postgis;"
```

Windows users can also execute `psql` from within the `c:\Program Files\PostgreSQL\9.3\bin\` folder as follows:

```
psql.exe -U postgres -d py_geoan_cb -c "CREATE EXTENSION postgis;"
```

8. Lastly, we'll create a schema called **geodata** to store our new spatial table. It is common to store spatial data in another schema outside the default `public` schema of PostgreSQL.

```
$ sudo -u postgres psql -d py_geoan_cb -c "CREATE SCHEMA geodata AUTHORIZATION pluto;"
```

Windows users can use the following command to do this:
```
psql.exe -U postgres -d py_geoan_cb -c "CREATE SCHEMA
geodata AUTHORIZATION pluto;"
```

How to do it...

1. Now, let's get into the actual importing of our Shapefile into a PostGIS database that will automatically create a new table from our Shapefile:

```python
#!/usr/bin/env python
# -*- coding: utf-8 -*-

import subprocess

# database options
db_schema = "SCHEMA=geodata"
overwrite_option = "OVERWRITE=YES"
geom_type = "MULTILINESTRING"
output_format = "PostgreSQL"

# database connection string
db_connection = """PG:host=localhost port=5432
  user=pluto dbname=py_test password=stars"""

# input shapefile
input_shp = "../geodata/bikeways.shp"

# call ogr2ogr from python
subprocess.call(["ogr2ogr","-lco", db_schema, "-lco",
overwrite_option,
  "-nlt", geom_type, "-f", output_format, db_connection,
input_shp])
```

2. Next, we'll call our script from the command line:

```
$ python ch03-01_shp2pg.py
```

How it works...

We begin with importing the standard Python `subprocess` module that will call the *ogr2ogr* command-line tool. Next, we'll set up a range of variables that are used as input arguments and provide various options for ogr2ogr to execute.

Starting with the PostgreSQL SCHEMA=geodata database, we set a nondefault database schema for the destination of our new table. It is a best practice to store your spatial data tables in a separate schema outside the public schema, which is the default. This practice will make backups and restores much easier and keeps your database better organized.

Next, we create a overwrite_option variable set to yes so that we can overwrite any table with the same name when it's created. This is helpful when you want to completely replace the table with new data; otherwise, it is recommended to use the -append option. We also specify the geometry type because, sometimes, ogr2ogr does not always guess the correct geometry type of our Shapefile so setting this value spares you that worry.

Now, we'll set our output_format variable with the PostgreSQL key word, telling ogr2ogr that we want to output data into a PostgreSQL database. This is then followed by the db_connection variable that specifies our database connection information. We must not forget that the database must already exist along with the geodata schema; otherwise, we will get an error.

The last input_shp variable is the full path to our Shapefile including the .shp file ending. We'll call the subprocess module and it will call the ogr2ogr command-line tool and pass along the variable options required to run the tool. We pass this function an array of arguments, starting with the first object in the array being the ogr2ogr command-line tool name. Following the name, we pass one option after another in the array to complete the call.

 Subprocess can be used to call any command-line tool directly. Subprocess takes a list of parameters separated by spaces. This passing of parameters is quite fussy, so make sure you follow along closely and don't add any extra spaces or commas.

Last but not least, we need to execute our script from the command line to actually import our Shapefile by calling the Python interpreter and passing the script. Now head over to the **PgAdmin** PostgreSQL database viewer and see if it's worked. Or, even better, open up Quantum GIS (www.qgis.org) and take a look at the newly created tables.

See also

If you would like to see the full list of options available with the ogr2ogr command, simply enter the following in the command line:

```
$ ogr2ogr -help
```

You will see the full list of options available. Also, visit http://gdal.org/ogr2ogr.html to read the available documentation.

For those of you who are curious to see how this call would run without using Python, the call directly to ogr2ogr will be as follows:

```
ogr2ogr -lco SCHEMA=geodata -nlt MULTILINE -f
"Postgresql" PG:"host=localhost port=5432 user=postgres
dbname=py_geoan_cb password=secret"
/home/mdiener/ch03/geodata/bikeways.shp
```

Batch importing a folder of Shapefiles into PostGIS using ogr2ogr

We would like to extend our last script to loop over a folder full of Shapefiles and import them into PostGIS. Most importing tasks involve more than one file to import, so this makes it a very practical task.

How to do it...

Our script will reuse the previous code in the form of a function so that we can batch process a list of Shapefiles to import into the PostgreSQL PostGIS database.

1. We will create our list of Shapefiles from a single folder for the sake of simplicity:

```python
#!/usr/bin/env python
# -*- coding: utf-8 -*-

import subprocess
import os
import ogr


def discover_geom_name(ogr_type):
    """

    :param ogr_type: ogr GetGeomType()
    :return: string geometry type name
    """
    return {ogr.wkbUnknown            : "UNKNOWN",
            ogr.wkbPoint              : "POINT",
            ogr.wkbLineString         : "LINESTRING",
            ogr.wkbPolygon            : "POLYGON",
            ogr.wkbMultiPoint         : "MULTIPOINT",
            ogr.wkbMultiLineString    : "MULTILINESTRING",
            ogr.wkbMultiPolygon       : "MULTIPOLYGON",
```

```
                ogr.wkbGeometryCollection : "GEOMETRYCOLLECTION",
                ogr.wkbNone               : "NONE",
                ogr.wkbLinearRing         :
"LINEARRING"}.get(ogr_type)


def run_shp2pg(input_shp):
    """

    input_shp is full path to shapefile including file
ending
    usage:  run_shp2pg('/home/geodata/myshape.shp')
    """


    db_schema = "SCHEMA=geodata"
    db_connection = """PG:host=localhost port=5432
                    user=pluto dbname=py_geoan_cb
password=stars"""
    output_format = "PostgreSQL"
    overwrite_option = "OVERWRITE=YES"
    shp_dataset = shp_driver.Open(input_shp)
    layer = shp_dataset.GetLayer(0)
    geometry_type = layer.GetLayerDefn().GetGeomType()
    geometry_name = discover_geom_name(geometry_type)
    print (geometry_name)

    subprocess.call(["ogr2ogr", "-lco", db_schema, "-lco",
overwrite_option,
                    "-nlt", geometry_name, "
-skipfailures",
                    "-f", output_format, db_connection,
input_shp])


# directory full of shapefiles
shapefile_dir = os.path.realpath('../geodata')


# define the ogr spatial driver type
shp_driver = ogr.GetDriverByName('ESRI Shapefile')


# empty list to hold names of all shapefils in directory
shapefile_list = []


for shp_file in os.listdir(shapefile_dir):
    if shp_file.endswith(".shp"):
        # apped join path to file name to outpout
"../geodata/myshape.shp"
```

```
            full_shapefile_path = os.path.join(shapefile_dir,
        shp_file)
            shapefile_list.append(full_shapefile_path)

    # loop over list of Shapefiles running our import function
    for each_shapefile in shapefile_list:
        run_shp2pg(each_shapefile)
        print ("importing Shapefile: " + each_shapefile)
```

2. Now, we can simply run our new script from the command line once again as follows:

   `$ python ch03-02_batch_shp2pg.py`

How it works...

Here, we are reusing our code from the previous script but have converted it into a Python function called `run_shp2pg (input_shp)` that takes exactly one argument and the complete path to the Shapefile that we want to import. The input argument must include the Shapefile ending, `.shp`.

We have a helper function that will get the geometry type as a string by reading in the Shapefile feature layer and outputting the geometry type so that the `ogr` commands know what to expect. This does not always work and some errors can occur. The `-skipfailures` option will plow over any errors that are thrown during insertion and will still populate our tables.

To begin with, we need to define the folder that contains all our Shapefiles to be imported. Next up, we can create an empty list object called `shapefile_list` that will hold the list of all our Shapefiles that we want to import.

The first `for` loop is used to get a list of all the Shapefiles in the directory specified using the standard Python `os.listdir()` function. We do not want all the files in this folder. We only want files with the file ending `.shp`; hence, the `if` statement that will evaluate to `True` if the file ends with `.shp`. Once the `.shp` file is found, we need to append the file path to the filename to create a single string that holds the path plus the Shapefile name and the `full_shapefile_path` variable. In the final part, we add each new file with its attached path to our `shapefile_list` list object so that we have our final list to loop through.

Now, it is time to loop through each Shapefile in our new list and run our `run_shp2pg(input_shp)` function for each Shapefile in the list, importing it into our PostgreSQL PostGIS database.

There's more...

If you have a lot of Shapefiles (and I mean a lot, as in 100 or more Shapefiles), performance will be one consideration and will, therefore, require a lot of machines with free resources.

Batch exporting a list of tables from PostGIS to Shapefiles

We will now change direction and take a look at how we can batch export a list of tables from our PostGIS database into a folder of Shapefiles. We'll again use the ogr2ogr command-line tool from within a Python script so that you can include it in your application programming work flow. Near the end, you can also see how all this works in one single command line.

How to do it...

1. The following script will fire the `ogr2ogr` command and loop over a list of tables to export the Shapefile format into an existing folder. So, let's take a look at how to do this as follows:

```python
#!/usr/bin/env python
# -*- coding: utf-8 -*-
#
import subprocess
import os

# folder to hold output Shapefiles
destination_dir = os.path.realpath('../geodata/temp')

# list of postGIS tables
postgis_tables_list = ["bikeways", "highest_mountains"]

# database connection parameters
db_connection = """PG:host=localhost port=5432 user=pluto
        dbname=py_geoan_cb password=stars active_schema=geodata"""

output_format = "ESRI Shapefile"

# check if destination directory exists
if not os.path.isdir(destination_dir):
    os.mkdir(destination_dir)
    for table in postgis_tables_list:
        subprocess.call(["ogr2ogr", "-f", output_format,
destination_dir,
                        db_connection, table])
        print("running ogr2ogr on table: " + table)
else:
    print("oh no your destination directory " +
destination_dir +
```

```
                       " already exist please remove it then run again")

     # commandline call without using python will look like this
     # ogr2ogr -f "ESRI Shapefile" mydatadump \
     # PG:"host=myhost user=myloginname dbname=mydbname
     password=mypassword" neighborhood parcels
```

2. Now, we'll call our script from the command line as follows:

```
$ python ch03-03_batch_postgis2shp.py
```

How it works...

Beginning with the simple import of our `subprocess` and `os` modules, we immediately define our destination directory where we want to store the exported Shapefiles. This variable is followed by the list of table names that we want to export. This list can only include files located in the same PostgreSQL schema. The schema is defined as the `active_schema` so that `ogr2ogr` knows where to find tables to export.

Once again, we define the output format as **ESRI Shapefile**. Now, we'll check whether the destination folder exists. If it does, we'll continue and call our loop. Then, we'll loop through the list of tables stored in our `postgis_tables_list` variable. If the destination folder does not exist, you will see an error printed on the screen.

There's more...

Programming an application and then executing the ogr2ogr command from inside your script is definitely quick and easy. On the other hand, for a one-off job, simply executing the command-line tool is what you want to do when exporting your list of Shapefiles. To do this in a one-liner, take a look at this information box.

> A one-line example of calling the ogr2ogr batch PostGIS table to Shapefiles is shown here if you simply want to execute this once and not in a scripting environment:
>
> **ogr2ogr -f "ESRI Shapefile" /home/ch03/geodata/temp PG:"host=localhost user=pluto dbname=py_geoan_cb password=stars" bikeways highest_mountains**
>
> The list of tables you want to export is located at the end as a list separated by spaces. The destination location of the exported Shapefiles is `../geodata/temp`. Note that this `/temp` directory must exist.

Converting an OpenStreetMap (OSM) XML to a Shapefile

OpenStreetMap (OSM) has a wealth of free data, but to use it with most other applications, we need to convert it to other formats, such as Shapefile or PostgreSQL PostGIS databases. This recipe will use the **ogr2ogr** tool to perform the conversion for us within a Python script. The benefit of this is, again, simplicity.

Getting ready

To get started, you will need to download the OSM data at `http://www.openstreetmap.org/export#map=17/37.80721/-122.47305` and save the file (`.osm`) to your `/ch03/geodata` directory. The download button is located on the left-hand side bar and, when pressed, it should immediately start the download (refer to the following screenshot). The area we are testing is in San Francisco, just before the **Golden Gate Bridge**.

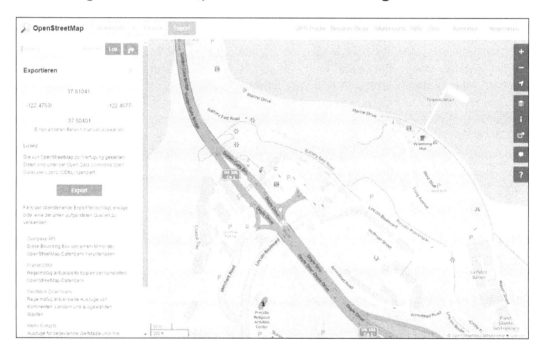

If you choose to download another area from OSM, feel free but make sure you take a small area similar to my example. If you select a larger area, the OSM web tool will give you a warning and disable the download button. The reason for this is simple: if the dataset is very large, it is most likely better suited for another tool, such as **osm2pgsql**, (http://wiki. openstreetmap.org/wiki/Osm2pgsql) for your conversion. If you need to get OSM data for a large area and want to export it to Shapefile, it would be advisable to use another tool, such as **osm2pgsql**, which will first import your data into a PostgreSQL database. Then, export the data from the PostGIS database to Shapefile using the **pgsql2shp** tool.

> A Python tool called **imposm** can be used to import the OSM data into a PostGIS database and is available at http://imposm.org/. Version 2 of it is written in Python and version 3 is written in the *go* programming language, if you want to give this a try as well.

How to do it...

Use the following steps to convert an OpenStreetMap (OSM) XML into a Shapefile:

1. Using the subprocess module, we will execute **ogr2ogr** to convert our OSM data that we downloaded into a new Shapefile:

```python
#!/usr/bin/env python
# -*- coding: utf-8 -*-

# convert / import osm xml .osm file into a Shapefile
import subprocess
import os
import shutil

# specify output format
output_format = "ESRI Shapefile"

# complete path to input OSM xml file .osm
input_osm = '../geodata/OSM_san_francisco_westbluff.osm'

# Windows users can uncomment these two lines if needed
# ogr2ogr = r"c:/OSGeo4W/bin/ogr2ogr.exe"
# ogr_info = r"c:/OSGeo4W/bin/ogrinfo.exe"

# view what geometry types are available in our OSM file
subprocess.call([ogr_info, input_osm])

destination_dir = os.path.realpath('../geodata/temp')
```

```
if os.path.isdir(destination_dir):
    # remove output folder if it exists
    shutil.rmtree(destination_dir)
    print("removing existing directory : " + destination_dir)
    # create new output folder
    os.mkdir(destination_dir)
    print("creating new directory : " + destination_dir)

    # list of geometry types to convert to Shapefile
    geom_types = ["lines", "points", "multilinestrings",
"multipolygons"]

    # create a new Shapefile for each geometry type
    for g_type in geom_types:

        subprocess.call([ogr2ogr,
                "-skipfailures", "-f", output_format,
                destination_dir, input_osm,
                "layer", g_type,
                "--config","OSM_USE_CUSTOM_INDEXING",
"NO"])
        print("done creating " + g_type)

# if you like to export to SPATIALITE from .osm
# subprocess.call([ogr2ogr, "-skipfailures", "-f",
#         "SQLITE", "-dsco", "SPATIALITE=YES",
#         "my2.sqlite", input_osm])
```

2. Now we can call our script from the command line:

   ```
   $ python ch03-04_osm2shp.py
   ```

Go and have a look in your `../geodata` folder to see the newly created Shapefiles and try to open them up in Quantum GIS, which is a free GIS software (`www.qgis.org`).

How it works...

This script should be clear as we are using the subprocess module call to fire our ogr2ogr command-line tool. We'll specify our OSM dataset as an input file, including the full path to the file. The Shapefile name is not supplied as ogr2ogr will output a set of Shapefiles, one for each geometry shape according to the geometry types it finds inside the OSM file. We only need to specify the name of the folder where we want ogr2ogr to export the Shapefiles to, automatically creating the folder if it does not exist.

Windows users: If you do not have your ogr2ogr tool mapped to your environment variables, you can simply uncomment the code at lines 16 and 17 and replace the path shown with the path on your machine to the Windows executables.

The first subprocess call prints out to the screen the geometry types found inside our OSM file. This is helpful in most cases to help identify what is available. Shapefiles can only support one geometry type per file, and this is why ogr2ogr outputs a folder full of Shapefiles, each one representing a separate geometry type.

Lastly, we call subprocess to execute ogr2ogr, passing in the output file type called ESRI Shapefile, the output folder, and the name of the OSM dataset.

Converting a Shapefile (vector) to a GeoTiff (raster)

Moving data from format to format also includes moving it from vector to raster or the other way round. In this recipe, we move data from a vector (Shapefile) to a raster (GeoTiff) with the Python `gdal` and `ogr` modules.

Getting ready

We need to be inside our virtual environment again, so fire it up so that we can access the `gdal` and `ogr` Python modules that we installed in *Chapter 1, Setting Up Your Geospatial Python Environment*.

As usual, enter your Python virtual environment with the `workon pygeoan_cb` command or this command:

```
$ source venvs/pygeoan_cb/bin/activate
```

A Shapefile is also needed, so be sure to download the source and access the `/ch03/geodata` folder (`https://github.com/mdiener21/python-geospatial-analysis-cookbook/archive/master.zip`).

How to do it...

Let's dive in and convert our golf course polygon Shapefile into a GeoTif; here comes the code:

1. Import the libraries `ogr` and `gdal`, and then define our output pixel size along with a value to assign to null:

    ```
    #!/usr/bin/env python
    # -*- coding: utf-8 -*-
    ```

```
from osgeo import ogr
from osgeo import gdal

# set pixel size
pixel_size = 1
no_data_value = -9999
```

2. Set up the input Shapefile we want to convert alongside the new GeoTiff raster that will be created when the script is executed:

```
# Shapefile input name
# input projection must be in Cartesian system in meters
# input wgs 84 or EPSG: 4326 will NOT work!!!
input_shp = r'../geodata/ply_golfcourse-strasslach3857.shp'

# TIF Raster file to be created
output_raster = r'../geodata/ply_golfcourse-strasslach.tif'
```

3. Now we need to create the input Shapefile object, get the layer information, and finally set the extent values:

```
# Open the data source get the layer object
# assign extent coordinates
open_shp = ogr.Open(input_shp)
shp_layer = open_shp.GetLayer()
x_min, x_max, y_min, y_max = shp_layer.GetExtent()
```

4. Here, we need to calculate the resolution distance to pixel value:

```
# calculate raster resolution
x_res = int((x_max - x_min) / pixel_size)
y_res = int((y_max - y_min) / pixel_size)
```

5. Our new raster type is a GeoTiff, so we must explicitly tell GDAL to get this driver. The driver is then able to create a new GeoTiff by passing in the filename or the new raster that we want to create, called the x direction resolution, followed by the y direction resolution, and then the number of bands; in this case, it is 1. Lastly, we set a new type of GDT_Byte raster:

```
# set the image type for export
image_type = 'GTiff'
driver = gdal.GetDriverByName(image_type)

new_raster = driver.Create(output_raster, x_res, y_res, 1,
gdal.GDT_Byte)
new_raster.SetGeoTransform((x_min, pixel_size, 0, y_max, 0,
-pixel_size))
```

6. Now we can access the new raster band and assign the no data values and the inner data values for the new raster. All the inner values will receive a value of 255 similar to what we set in the `burn_values` variable:

```
# get the raster band we want to export too
raster_band = new_raster.GetRasterBand(1)

# assign the no data value to empty cells
raster_band.SetNoDataValue(no_data_value)

# run vector to raster on new raster with input Shapefile
gdal.RasterizeLayer(new_raster, [1], shp_layer,
burn_values=[255])
```

7. Here we go; let's run this script to see what our new raster looks like:

```
$ python ch03-05_shp2raster.py
```

Our resulting raster should look like what is shown in the following screenshot if you open it using **QGIS** (`http://www.qgis.org`):

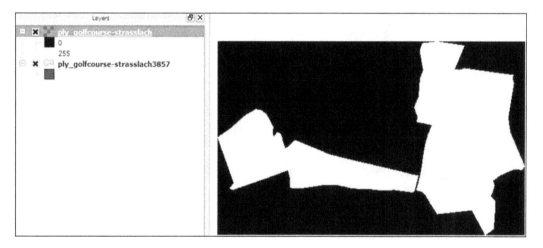

How it works...

There are several steps involved in this code so follow along as some points could lead to trouble if you are not sure what values to input. We start with the import of the *gdal* and *ogr* modules, respectively, since they will do the work for us by inputting a Shapefile (vector) and outputting a GeoTiff (raster).

The `pixel_size` variable is very important since it will determine the size of the new raster that we will create. In this example, we only have two polygons, so we set `pixel_size = 1` to keep a fine border between them. If you have many polygons stretching across the globe in one Shapefile, it is wiser to set this value to 25 or more. Otherwise, you could end up with a 10 GB raster and your machine will run all night long! The `no_data_value` is needed to tell GDAL what values to set in the empty space around our input polygons and we set it to `-9999` for easy identification.

Next, we simply set the input Shapefile stored in EPSG:3857 Web Mercator and output GeoTiff. Check to make sure that you change the filenames accordingly if you want to use some other dataset. We start by working with the OGR module to open the Shapefile and retrieve its layer information and the extent information. The extent is important because it is used to calculate the size of the output raster width and height values that must be integers represented by the `x_res` and `y_res` variables.

Note that the projection of your Shapefile must be in meters and not degrees. This is very important since this will NOT work in EPSG:4326, WGS 84, for example. The reason for this is that the coordinate units are LAT/LON. This means that WGS84 is not a flat plane projection and cannot be drawn as is. Our `x_res` and `y_res` values would evaluate to 0 since we cannot get a real ratio using degrees. This is a result of us not being able to simply subtract coordinate *x* from coordinate *y* because the units are in degrees and not in a flat plane meters projection.

Now, moving on to the raster setup, we define the type of raster we want to export as a `Gtiff`. Then, we'll get the correct GDAL driver by the raster type. Once the raster type is set, we can create a new empty raster dataset, passing in a raster filename, the width, the height of the raster in pixels, the number of raster bands, and finally, the type of rasters in GDAL terms, such as `gdal.GDT_Byte`. These five parameters are mandatory to create a new raster.

Next, we call `SetGeoTransform` that handles transforming between pixel/line raster space and projection coordinate space. We'll want to activate `band 1` as it is the only band we have in our raster. Then, we'll assign the no data value to all our empty space around a polygon.

The final step is to call the `gdal.RasterizeLayer()` function and pass in our new raster, band, Shapefile, and the value to assign to the inside of our raster. All the pixels inside the polygon will be assigned a value of 255.

See also

If you are interested, you can visit the `gdal_rasterize` command-line tool at `http://www.gdal.org/gdal_rasterize.html`. You can run this straight from the command line.

Converting a raster (GeoTiff) to a vector (Shapefile) using GDAL

We have now looked at how we can go from a vector to a raster, so it is now time to go from a raster to a vector. This method is much more common because most of our vector data is derived from remotely sensed data, such as satellite images, orthophotos, or some other remote sensing dataset, such as `lidar`.

Getting ready

As usual, enter the `workon pygeoan_cb` command in your Python virtual environment:

```
$ source venvs/pygeoan_cb/bin/activate
```

How to do it...

This recipe only requires four steps utilizing OGR and GDAL so please open up a new file for your code:

1. Import the `ogr` and `gdal` modules and go straight ahead and open the raster we want to convert by passing it the filename on disk and getting a raster band:

```python
#!/usr/bin/env python
# -*- coding: utf-8 -*-

from osgeo import ogr
from osgeo import gdal

#  get raster data source
open_image = gdal.Open( "../geodata/cadaster_borders-2tone-
black-white.png" )
input_band = open_image.GetRasterBand(3)
```

2. Set up the output vector file as a Shapefile with output_shp, and then get a Shapefile driver. Now, we can create the output from our driver and create a layer as follows:

```python
#  create output data source
output_shp = "../geodata/cadaster_raster"
shp_driver = ogr.GetDriverByName("ESRI Shapefile")

# create output file name
output_shapefile = shp_driver.CreateDataSource( output_shp
+ ".shp" )
new_shapefile = output_shapefile.CreateLayer(output_shp,
srs = None )
```

3. The final step is to run the `gdal.Polygonize` function that does the heavy lifting by converting our raster to a vector as follows:

```
gdal.Polygonize(input_band, None, new_shapefile, -1, [],
callback=None)
new_shapefile.SyncToDisk()
```

4. Execute the new script as follows:

```
$ python ch03-06_raster2shp.py
```

How it works...

Working with `ogr` and `gdal` is similar in all our recipes; we must define the inputs and get an appropriate file driver to open the files. The GDAL library is very powerful and in only one line of code we can convert a raster to a vector with the help of the `gdal.Polygonize` function. The preceding code is simply setup code to define which format we want to work with and then set up an appropriate driver to input and output our new file.

Creating a Shapefile from point data stored in Microsoft Excel

Excel files are so common these days that often an analyst or developer receives an Excel file that needs to be mapped out. Sure, we could save these to a `.csv` file and then use the great Python standard *csv* module but this involves an extra manual step. We will take a look at how to read a very simple Excel file that contains a list of Europe's highest mountains. This data set is derived from `http://www.geonames.org`.

Getting ready

We are going to need one new Python library to read a Microsoft Excel file and this library is **xlrd** (`http://www.python-excel.org`).

 This library can only READ an Excel file; if you are looking to write out to an Excel file, download and install **xlwt**.

First, fire up you virtual environment from your `workon pygeoan_cb` Linux machine, run `pip install xlrd`, and you are off to the races.

To write out to a new Shapefile, we will use the pyshp library we installed in *Chapter 1, Setting Up Your Geospatial Python environment*, so that there is no need to do anything.

The data is located in your downloads in `/ch03/geodata` and the output Shapefile will also be written to this location after you go through this recipe.

How to do it...

So let's get started with some code:

1. Start with the import of `xlrd` and the pyshp module; note that the import name is `shapefile` and not pyshp as the module name would imply:

```
#!/usr/bin/env python
# -*- coding: utf-8 -*-

import xlrd
import shapefile
```

2. Open the Excel file using the xlrd module and create a variable to hold the Excel sheet. We reference the first sheet in the Excel file by an index number, always starting with (0) in the first sheet:

```
excel_file = xlrd.open_workbook("../geodata/
highest-mountains-europe.xlsx")

# get the first sheet
sh = excel_file.sheet_by_index(0)
```

3. Create the Shapefile object as follows:

```
w = shapefile.Writer(shapefile.POINT)
```

4. Define the new Shapefile fields and their data types. *F* stands for float and *C* is for character:

```
w.field('GeoNameId','F')
w.field('Name', 'C')
w.field('Country', 'C')
w.field('Latitude', 'F')
w.field('Longitude', 'F')
w.field('Altitude', 'F')
```

5. Loop through each row in the Excel file and create the geometry values along with attributes:

```
for row_number in range(sh.nrows):
    # skips over the first row since it is the header row
    if row_number == 0:
        continue
    else:
        x_coord = sh.cell_value(rowx=row_number, colx=4)
        y_coord = sh.cell_value(rowx=row_number, colx=3)
        w.point(x_coord, y_coord)
```

```
        w.record(GeoNameId=sh.cell_value(rowx=row_number,
colx=0), Name=sh.cell_value(rowx=row_number, colx=1),
            Country=sh.cell_value(rowx=row_number,
colx=2), Latitude=sh.cell_value(rowx=row_number, colx=3),
            Longitude=sh.cell_value(rowx=row_number,
colx=4),Altitude=sh.cell_value(rowx=row_number, colx=5))
        print "Adding row: " + str(row_number) + " creating
mount: " + sh.cell_value(rowx=row_number, colx=1)
```

6. Lastly, we'll create the new Shapefile in the `/ch03/geodata` folder as follows:

```
w.save('../geodata/highest-mountains')
```

7. Go ahead and execute our new `ch03-07_excel2shp.py` script from the command line as follows:

```
$ python ch03-07_excel2shp.py
```

How it works...

The Python code reads similar to making a description of how code works and is almost all too easy to explain. We start with importing our new *xlrd* module along with the Shapefile module needed to write out to a Shapefile. Taking a look at our Excel file, we see which fields are available and locate where the *X* coordinate (longitude) and *Y* coordinate (latitude) are positioned. This position index number remembers the starting point by counting from 0 for the first column.

Our Excel file also has a header row and this is, of course, not to be included in the new data attributes; this is why we check to see whether row numbers are equal to 0—that is, the first row—and then continue. The continue statements allow the code to continue without an error and enter the `else` statement where we define the index positions of our columns. Each column is referenced using the `pyshp` syntax, referencing the columns by name to make the code even easier to read.

We call the `w.point` pyshp function to create the point geometry passing in our x and y coordinates as floats. The `xlrd` module converts the values for us automatically into floats, which is nice. All we need to do in the end is use the pyshp save function to write out to our `/ch03/geodata` folder. There is no need to add the `.shp` extension; pyshp handles this for us and outputs `.shp`, `.dbf`, and `.shx`.

Note that a `.prj` projection file is not automatically output. If you would like to have the projection information exported as well, you will need to manually create it like this:

```
# create the PRJ file
filename = 'highest-mountains'
prj = open("%s.prj" % filename, "w")
epsg = 'GEOGCS["WGS 84",DATUM["WGS_1984",SPHEROID[
"WGS 84",6378137,298.257223563]],PRIMEM["Greenwich
",0],UNIT["degree",0.0174532925199433]]'
prj.write(epsg)
prj.close()
```

Converting an ESRI ASCII DEM to an image height map

To end this chapter with a bang, here is the most complicated conversion we have seen so far and the most fun as well. Input is an elevation dataset that's stored in *ASCII* format, more specifically, Arc/Info ASCII Grid, for short with the AAIGrid with the (`.asc`) file ending. Our output is a *heightmap* image (`http://en.wikipedia.org/wiki/Heightmap`). A heightmap image is an image that stores height elevation as a pixel value. A heightmap is also simply known as a **digital elevation model** (**DEM**). The benefit of using an image to store elevation data is that it is *web compatible* and we can use this in a 3D visualization with **threejs**, for example, as shown in *Chapter 10, Visualizing Your Analysis*.

We need to be careful with regard to the output image format because simply storing an 8-bit image limits us to only storing 0 to 255 height values, which is typically not enough. The output image should store a minimum of 16-bits, giving us a range from -32,767 to 32,767. If I am correct, the tallest mountain on earth is Mt. Everest at a height of 8,848 m, so a 16-bit image should be more than enough to hold our elevation data.

Getting ready

A DEM is needed to run this exercise so please make sure you have downloaded the code and geodata included at `https://github.com/mdiener21/python-geospatial-analysis-cookbook/archive/master.zip` and download the sample DEM needed to process. You do not need to run your script from within your virtual environment because this script will be executing standard Python modules and several GDAL built-in tools installed with GDAL. This simply means that you need to make sure your GDAL utilities are properly installed and running on your machine. (Refer to *Chapter 2, Working with Projections*, for the reference installation.)

How to do it...

We will execute this script by calling several GDAL utility scripts installed by `gdal` from our Python script:

1. We'll start by importing the subprocess standard module; this will be used to execute our GDAL utility functions. Then, we'll set the base path to where we will store our geodata for input files, temporary files, and output files:

```
#!/usr/bin/env python
# -*- coding: utf-8 -*-
import subprocess
from osgeo import gdal

path_base = "../geodata/"
```

2. Windows users who have installed GDAL using the great OSGeo4w installer might want to specify the path directly to the GDAL utilities if it is not available in the Windows Environment variables as follows:

```
# gdal_translate converts raster data between different
formats

command_gdal_translate =
"c:/OSGeo4W/bin/gdal_translate.exe"
command_gdalinfo = "c:/OSGeo4W/bin/gdalinfo.exe"
```

3. Linux users can use these variables:

```
command_gdal_translate = "gdal_translate"
command_gdalinfo = "gdalinfo"
command_gdaldem = "gdaldem"
```

4. We'll create a set of variables to hold our input DEM, output files, temporary files, and our final output file. The variables concatenate the base path folder to the filename as follows:

```
orig_dem_asc = path_base + "original_dem.asc"

temp_tiff = path_base + "temp_image.tif"

output_envi = path_base + "final_envi.bin"
```

5. Then, we'll call the `gdal_translate` command to create our new temporary GeoTiff as follows:

```
# transform dem to tiff
dem2tiff = command_gdal_translate + " " + orig_dem_asc + "
" + temp_tiff
print ("now executing this command: " + dem2tiff)
subprocess.call(dem2tiff.split(), shell=False)
```

6. Next, we'll open the temp GeoTiff and read the information about the tiff to find out the minimum and maximum height values stored in our data. This is not needed to complete the script but is very useful to identify your maximum and minimum height values:

```
ds = gdal.Open(temp_tiff, gdal.GA_ReadOnly)
band = ds.GetRasterBand(1)
print 'Band Type=', gdal.GetDataTypeName(band.DataType)
min = band.GetMinimum()
max = band.GetMaximum()
if min is None or max is None:
    (min, max) = band.ComputeRasterMinMax(1)
print 'Min=%.3f, Max=%.3f' % (min, max)
min_elevation = str(int(round(min)))
max_elevation = str(int(round(max)))
```

7. Then, call the gdal_translate utility with the following parameters, setting the scale range from its original min/max values to a new scale ranging from the 0 to 65,535 values. Specify the -ot output type to be in the vENVI format using our temporary GeoTiff as the input:

```
tif_2_envi = command_gdal_translate + " -scale -ot UInt16
-outsize 500 500 -of ENVI " \
             + temp_tiff + " " + output_envi
```

8. Let's run our new ch03-08_dem2heightmap.py script from the command line:

   subprocess.call(tif_2_envi.split(),shell=False)

9. Let's run our new ch03-08_dem2heightmap.py script from the command line:

   python ch03-08_dem2heightmap.py

The result is that you have a new .bin file located in your /ch03/geodata/ folder that stores your new ENVI 16-bit image including all your elevation data. The image height map can now be used in your 3D software, such as Blender (www.blender.org), Unity (www.unity3d.com), or in an even cooler web application using a JavaScript library such as threejs.

How it works...

Let's start with the imports, and then we'll specify the base path to where our inputs and outputs will be stored. After this, we'll see the actual commands we used to execute the gdal_translate transformation. The commands for Windows and Linux are for you to decide whether to use or not and this depends on how you have set up up your machine. We then set our variable to define the input DEM, temporary GeoTiff, and the output ENVI height map image.

At last, we can call the first transformation that converts our DEM ASCII file into a GeoTiff with the `gdal_translate` utility. Now to get a little information about our data, we print out the `min` and `max` height values to the screen. Sometimes, this is very useful when transforming, allowing you to check whether the output data actually contains the input height values and that nothing went astray during conversion.

In the end, we simply call the `gdal_translate` utility once again to convert our GeoTiff into an ENVI heightmap image. The -scale with no parameters automatically fills our 16-bit image with values ranging from 0 to 65,535. Our next parameter is `-ot`, which specifies the output type as 16-bit followed by `-outsize 500 500`, setting the output image size to 500 x 500 pixels. Lastly, `-of ENVI` is our output format followed by the name of the input GeoTiff and the name of the output height map.

A typical work flow when working with DEM's is as follows:

1. Download a DEM that is usually a very large file and covers a large geographic region.
2. Clip the DEM to a smaller region of interest.
3. Convert the clipped region to another format.
4. Export the DEM as a heightmap image.

We introduce `.split()` that will return a Python list of words, separated by a character. In our case, the separator character is a *single space* character but you could split based on any other character or a combination of characters (refer to the Python documentation at `https://docs.python.org/2/library/string.html#string.split`) This helps us reduce the amount of concatenating that we need to do in our code.

4

Working with PostGIS

In this chapter we will cover the following topics:

- ▶ Executing a PostGIS ST_Buffer analysis query and exporting it to GeoJSON
- ▶ Finding out whether a point is inside a polygon
- ▶ Splitting LineStrings at intersections using ST_Node
- ▶ Checking the validity of LineStrings
- ▶ Executing a spatial join and assigning point attributes to a polygon
- ▶ Conducting a complex spatial analysis query using ST_Distance()

Introduction

A spatial database is nothing but a standard database that can store geometry and execute spatial queries in their simplest forms. We will explore how to run spatial analysis queries, handle connections, and more, all from our Python code. Your ability to answer spatial questions such as "I want to locate all the hotels that are within 2 km of a golf course and less than 5 km from a park" is where PostGIS comes into play. This chaining of requests into a model is where the powers of spatial analysis shine.

We will work with the most popular and powerful open source spatial database called **PostgreSQL**, along with the **PostGIS** extension, including over 150 functions. Basically, we'll get a full-blown GIS with complex spatial analysis functions for both vectors and rasters, spatial data types, and diverse methods to move spatial data around.

If you are looking for more information on PostGIS and a good read, please check out *PostGIS Cookbook* by *Paolo Corti* (available at `https://www.packtpub.com/big-data-and-business-intelligence/postgis-cookbook`). This book explores the wider use of PostGIS and includes a full chapter on PostGIS Programming using Python.

Executing a PostGIS ST_Buffer analysis query and exporting it to GeoJSON

Let's start by executing our first spatial analysis query from Python against our already running PostgreSQL and PostGIS database. The goal is to generate a 100 m buffer around all schools and export the new buffer polygon to GeoJSON, including the name of a school. The end result will be shown on this map, available (`https://github.com/mdiener21/python-geospatial-analysis-cookbook/blob/master/ch04/geodata/out_buff_100m.geojson`) on GitHub.

Quick visualizations of GeoJSON data using GitHub is a fast and simple way to create a web map without coding a single line. Note that the data is then free for everyone else to download if you are using a public and free GitHub account. Private GitHub accounts mean the data, that is, GeoJSON, will also remain private if data privacy or sensitivity is an issue.

Getting ready

To get started, we'll use our data in the PostGIS database. We will begin by accessing our `schools` table that we uploaded to PostGIS in the Batch importing a folder of Shapefiles into PostGIS using ogr2ogr recipe of *Chapter 3, Moving Spatial Data from One Format to Another*.

Connecting to a PostgreSQL and PostGIS database is accomplished with **Psycopg**, which is a Python DB API (http://initd.org/psycopg/) implementation. We've already installed this in *Chapter 1, Setting Up Your Geospatial Python Environment* along with PostgreSQL, Django, and PostGIS.

For all the following recipes, enter your virtual environment, pygeoan_cb, so that you have access to your libraries using this command:

```
workon pygeoan_cb
```

How to do it...

1. The long road is not so long after all, so follow along:

```python
#!/usr/bin/env python
# -*- coding: utf-8 -*-

import psycopg2
import json
from geojson import loads, Feature, FeatureCollection

# NOTE change the password and username
# Database Connection Info
db_host = "localhost"
db_user = "pluto"
db_passwd = "stars"
db_database = "py_geoan_cb"
db_port = "5432"

# connect to DB
conn = psycopg2.connect(host=db_host, user=db_user,
        port=db_port, password=db_passwd,
database=db_database)

# create a cursor
cur = conn.cursor()

# the PostGIS buffer query
buffer_query = """SELECT ST_AsGeoJSON(ST_Transform(
        ST_Buffer(wkb_geometry, 100,'quad_segs=8'),4326))
        AS geom, name
        FROM geodata.schools"""

# execute the query
cur.execute(buffer_query)
```

```
# return all the rows, we expect more than one
dbRows = cur.fetchall()

# an empty list to hold each feature of our feature collection
new_geom_collection = []

# loop through each row in result query set and add to my feature
collection
# assign name field to the GeoJSON properties
for each_poly in dbRows:
    geom = each_poly[0]
    name = each_poly[1]
    geoj_geom = loads(geom)
    myfeat = Feature(geometry=geoj_geom,
properties={'name': name})
    new_geom_collection.append(myfeat)

# use the geojson module to create the final Feature
Collection of features created from for loop above
my_geojson = FeatureCollection(new_geom_collection)

# define the output folder and GeoJSon file name
output_geojson_buf = "../geodata/out_buff_100m.geojson"


# save geojson to a file in our geodata folder
def write_geojson():
    fo = open(output_geojson_buf, "w")
    fo.write(json.dumps(my_geojson))
    fo.close()

# run the write function to actually create the GeoJSON
file
write_geojson()

# close cursor
cur.close()

# close connection
conn.close()
```

How it works...

The database connection is using the `pyscopg2` module, so we import the libraries at the start alongside `geojson` and the standard `json` modules to handle our GeoJSON export.

Our connection is created and then followed immediately with our SQL Buffer query string. The query uses three PostGIS functions. Working your way from the inside out, you will see the `ST_Buffer` function taking in the geometry of the school points followed by the 100 m buffer distance and the number of circle segments that we would like to generate. `ST_Transform` then takes the newly created buffer geometry and transforms it into the WGS84 coordinate system (EPSG: 4326) so that we can display it on GitHub, which only displays WGS84 and the projected GeoJSON. Lastly, we'll use the `ST_asGeoJSON` function to export our geometry as the GeoJSON geometry.

 PostGIS does not export the complete GeoJSON syntax, only the geometry in the form of the GeoJSON geometry. This is the reason that we need to complete our GeoJSON using the Python `geojson` module.

All of this means that we not only perform analysis on the query, but we also specify the output format and coordinate system all in one go.

Next, we will execute the query and fetch all the returned objects using `cur.fetchall()` so that we can later loop through each returned buffer polygon. Our `new_geom_collection` list will store each of the new geometries and the feature names. Next, in the `for` loop function, we'll use the `geojson` module function, `loads(geom)`, to input our geometry into a GeoJSON geometry object. This is followed by the `Feature()` function that actually creates our GeoJSON feature. This is then used as the input for the `FeatureCollection` function where the final, completed GeoJSON is created.

Lastly, we'll need to write this new GeoJSON file to disk and save it. Hence, we'll use the new file object where we use the standard Python `json.dumps` module to export our `FeatureCollection`.

We'll do a little clean up to close the cursor object and connection. Bingo! We are now done and can visualize our final results.

Finding out whether a point is inside a polygon

A point inside a polygon analysis query is a very common spatial operation. This query can identify objects located within an area such as a polygon. The area of interest in this example is a 100 m buffer polygon around bike paths and we would like to locate all schools that are inside this polygon.

Getting ready

In the previous section, we used the `schools` table to create a buffer. This time around, we will use this table as our input points table. The `bikeways` table that we imported in *Chapter 3*, *Moving Spatial Data from One Format to Another*, will be used as our input lines to generate a new 100 m buffer polygon. Be sure, however, that you have the two datasets in your local PostgreSQL database.

How to do it...

1. Now, let's dive into some more code to find schools located within 100 m of the bikeways in order to find points inside a polygon:

```python
#!/usr/bin/env python
# -*- coding: utf-8 -*-

import json
import psycopg2
from geojson import loads, Feature, FeatureCollection


# Database Connection Info
db_host = "localhost"
db_user = "pluto"
db_passwd = "stars"
db_database = "py_geoan_cb"
db_port = "5432"

# connect to DB
conn = psycopg2.connect(host=db_host, user=db_user,
port=db_port, password=db_passwd, database=db_database)

# create a cursor
cur = conn.cursor()

# uncomment if needed
# cur.execute("Drop table if exists geodata.bikepath_100m_buff;")

# query to create a new polygon 100m around the bikepath
new_bike_buff_100m = """ CREATE TABLE
geodata.bikepath_100m_buff
        AS SELECT name,
        ST_Buffer(wkb_geometry, 100) AS geom
        FROM geodata.bikeways; """
```

```python
# run the query
cur.execute(new_bike_buff_100m)

# commit query to database
conn.commit()

# query to select schools inside the polygon and output
geojson
is_inside_query = """ SELECT s.name AS name,
    ST_AsGeoJSON(ST_Transform(s.wkb_geometry,4326)) AS geom
    FROM geodata.schools AS s,
    geodata.bikepath_100m_buff AS bp
        WHERE ST_WITHIN(s.wkb_geometry, bp.geom); """

# execute the query
cur.execute(is_inside_query)

# return all the rows, we expect more than one
db_rows = cur.fetchall()

# an empty list to hold each feature of our feature
collection
new_geom_collection = []

def export2geojson(query_result):
    """
    loop through each row in result query set and add to my
feature collection
    assign name field to the GeoJSON properties
    :param query_result: pg query set of geometries
    :return: new geojson file
    """

    for row in db_rows:
        name = row[0]
        geom = row[1]
        geoj_geom = loads(geom)
        myfeat = Feature(geometry=geoj_geom,
                    properties={'name': name})
        new_geom_collection.append(myfeat)

    # use the geojson module to create the final Feature
    # Collection of features created from for loop above
    my_geojson = FeatureCollection(new_geom_collection)
```

```
# define the output folder and GeoJSon file name
output_geojson_buf =
"../geodata/out_schools_in_100m.geojson"

# save geojson to a file in our geodata folder
def write_geojson():
    fo = open(output_geojson_buf, "w")
    fo.write(json.dumps(my_geojson))
    fo.close()

# run the write function to actually create the GeoJSON
file
    write_geojson()


export2geojson(db_rows)
```

You can now view your newly created GeoJSON file on a great little site created by Mapbox at `http://www.geojson.io`. Simply drag and drop your GeoJSON file from Windows Explorer in Windows or Nautilus in Ubuntu onto the `http://www.geojson.io` web page and, Bob's your uncle, you should see 50 or so schools that are located within 100 m of a bikeway in Vancouver.

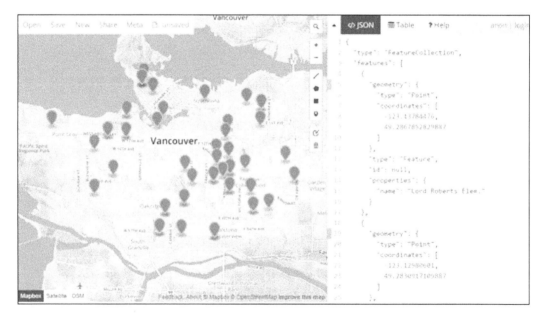

How it works...

We will reuse code to make our database connection, so this should be familiar to you at this point. The `new_bike_buff_100m` query string contains our query to generate a new 100 m buffer polygon around all the bikeways. We need to execute this query and commit it to the database so that we can access this new set of polygons as input to our actual query that will find schools (points) located inside this new buffer polygon.

The `is_inside_query` string actually does the hard work for us by selecting selecting the values from the field `name` and the geometry from the `geom` field. The geometry is wrapped up in two other PostGIS functions to allow us to export our data as GeoJSON in the WGS 84 coordinate system. This will be the input geometry needed to generate our final new GeoJSON file.

The `WHERE` clause uses the `ST_Within` function to see whether a point is inside the polygon and returns `True` if the point is within the buffer polygon.

Now, we've created a new function that simply wraps up our export to the GeoJSON code that was used in the previous, *Executing a PostGIS ST_Buffer analysis query and exporting it to GeoJSON*, recipe. This new `export2geojson` function simply takes one input of our PostGIS query and outputs a GeoJSON file. To set the name and location of the new output file, simply replace the path and name within the function.

Finally, all we need to do is call the new function to export the GeoJSON file using the `db_rows` variable that contains our list of schools as points located within the 100 m buffer polygon.

There's more...

This example to find all schools located within 100 m of the bike paths could be completed using another PostGIS function called `ST_Dwithin`.

The SQL to select all the schools located within 100 m of the bikepaths would look like this:

```
SELECT *  FROM geodata.bikeways as b, geodata.schools as s where ST_
DWithin(b.wkb_geometry, s.wkb_geometry, 100)
```

Splitting LineStrings at intersections using ST_Node

Working with road data is usually a tricky business because the validity of the data and data structure plays a very important role. If you want to do anything useful with your road data, such as building a routing network, you will need to prepare the data first. The first task is usually to segmentize your lines, which means splitting all lines at intersections where LineStrings cross each other, creating a base network road dataset.

> Be aware that this recipe will split all lines on all intersections regardless of whether, for example, there is a road-bridge overpass where no intersection should be created.

Getting ready

Before we get into the details of how to do this, we will use a small section of the **OpenStreetMap** (**OSM**) road data for our example. The OSM data is available in your /ch04/geodata/ folder called vancouver-osm-data.osm. This data was simply downloaded from the www. openstreetmap.org home page using the **Export** button located at the top of the page:

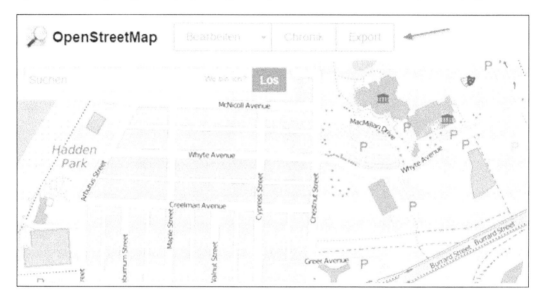

The OSM data contains not only roads but all the other points and polygons located within the extent that I have chosen. The region of interest is again the Burrard Street bridge in Vancouver.

We are going to need to extract all the roads and import them into our PostGIS table. This time, let's try using the ogr2ogr command line directly from the console to upload the OSM streets to our PostGIS database:

```
ogr2ogr -lco SCHEMA=geodata -nlt LINESTRING -f "PostgreSQL"
PG:"host=localhost port=5432 user=pluto dbname=py_geoan_cb
password=stars" ../geodata/vancouver-osm-data.osm lines -t_srs EPSG:3857
```

This assumes that your OSM data is in the /ch04/geodata folder and the command is run while you are located in the /ch04/code folder.

Now this really long thing means that we connect to our PostGIS database as our output and input the `vancouver-osm-data.osm` file. Create a new table called `lines` and transform the input OSM projection to EPSG:3857. All data exported from OSM is in EPSG:4326. You can, of course, leave it in this system and simply remove the `-t_srs EPSG:3857` part of the command line option.

We are now ready to rock and roll with the splitting at intersections. If you like, go ahead and open the data in **QGIS** (**Quantum GIS**). In QGIS, you will see that the road data is not split at all road intersections as shown in this screenshot:

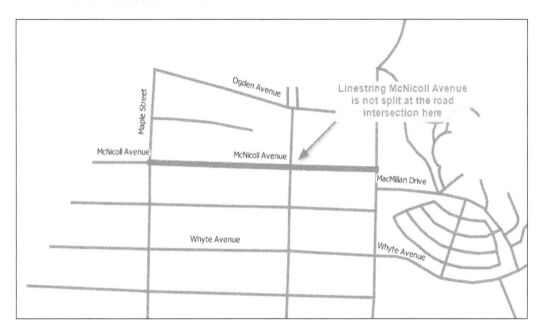

Here, you can see that **McNicoll Avenue** is a single LineString crossing over **Cypress Street**. After we've completed the recipe, we will see that **McNicoll Avenue** will be split at this intersection.

How to do it...

1. Running through the Python code is quite straightforward since the hard work is done in one SQL query. So follow along:

```python
#!/usr/bin/env python
# -*- coding: utf-8 -*-

import psycopg2
import json
from geojson import loads, Feature, FeatureCollection
```

```python
# Database Connection Info
db_host = "localhost"
db_user = "pluto"
db_passwd = "stars"
db_database = "py_geoan_cb"
db_port = "5432"

# connect to DB
conn = psycopg2.connect(host=db_host, user=db_user,
    port=db_port, password=db_passwd, database=db_database)

# create a cursor
cur = conn.cursor()

# drop table if exists
# cur.execute("DROP TABLE IF EXISTS geodata.split_roads;")

# split lines at intersections query
split_lines_query = """
 CREATE TABLE geodata.split_roads
    (ST_Node(ST_Collect(wkb_geometry)))).geom AS geom
    FROM geodata.lines;"""

cur.execute(split_lines_query)
conn.commit()

cur.execute("ALTER TABLE geodata.split_roads ADD COLUMN id
serial;")
cur.execute("ALTER TABLE geodata.split_roads ADD CONSTRAINT split_
roads_pkey PRIMARY KEY (id);")

# close cursor
cur.close()

# close connection
conn.close()
```

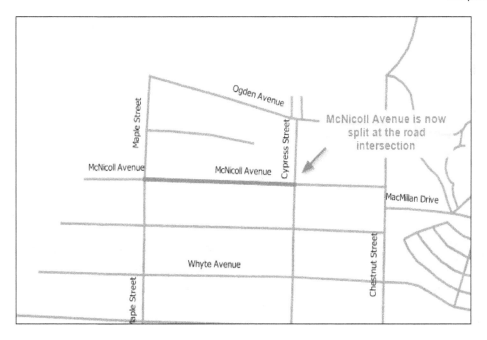

Well, this was quite simple and we can now see that **McNicoll Avenue** is split at the intersection with **Cypress Street**.

How it works...

Looking at the code, we can see that the database connection remains the same and the only new thing is the query itself that creates the intersection. Here three separate PostGIS functions are used to obtain our results:

- ▶ The first function, when working our way inside-out in the query, starts with `ST_Collect(wkb_geometry)`. This simply takes our original geometry column as input. The simple combining of the geometries is all that is going on here.

- ▶ Next up is the actual splitting of the lines using the `ST_Node(geometry)`, inputting the new geometry collection and nodding, which splits our LineStrings at intersections.

- ▶ Finally, we'll use `ST_Dump()` as a set returning function. This means that it basically explodes all the LineString geometry collections into individual LineStrings. The end of the query with `.geom` specifies that we only want to export the geometry and not the returned array numbers of the split geometry.

Now, we'll execute and commit the query to the database. The commit is an important part because, otherwise, the query will be run but it will not actually create the new table that we are looking to generate. Last but not least, we can close down our cursor and connection. That is that; we now have split LineStrings.

Be aware that the new split LineStrings do NOT contain the street names and other attributes. To export the names, we would need to do a join on our data. Such a query to include the attributes on the newly created LineStrings could look like this:

```
CREATE TABLE geodata.split_roads_attributes AS SELECT
    r.geom,
    li.name,
    li.highway
FROM
    geodata.lines li,
    geodata.split_roads r
WHERE
    ST_CoveredBy(r.geom, li.wkb_geometry)
```

Checking the validity of LineStrings

Working with road data has many areas to watch out for and one of these is invalid geometry. Our source data is OSM and is, therefore, collected by a community of users that are not trained by GIS professionals, resulting in errors. To execute spatial queries, the data must be valid or we will have results with errors or no results at all.

PostGIS includes the ST_isValid() function that returns True/False on the basis of whether a geometry is valid or not. There is also the ST_isValidReason() function that will output a text description of the geometry error. Finally, the ST_isValidDetail() function will return if the geometry is valid along with the reason and location of the geometry error. These three functions all accomplish similar tasks and selecting one depends on what you want to accomplish.

How to do it...

1. Now, to determine if geodata.lines are valid, we will run another query that will list all invalid geometries if there are any:

```
#!/usr/bin/env python
# -*- coding: utf-8 -*-

import psycopg2

# Database Connection Info
db_host = "localhost"
db_user = "pluto"
db_passwd = "stars"
db_database = "py_geoan_cb"
db_port = "5432"
```

```
# connect to DB
conn = psycopg2.connect(host=db_host, user=db_user,
    port=db_port, password=db_passwd, database=db_database)

# create a cursor
cur = conn.cursor()

# the PostGIS buffer query
valid_query = """SELECT
                ogc_fid,
                ST_IsValidDetail(wkb_geometry)
            FROM
                geodata.lines
            WHERE NOT
                ST_IsValid(wkb_geometry);
            """

# execute the query
cur.execute(valid_query)

# return all the rows, we expect more than one
validity_results = cur.fetchall()

print validity_results

# close cursor
cur.close()

# close connection
conn.close();
```

This query should return an empty Python list, which means that we have no invalid geometries. If there are objects in your list, then you'll know that you have some manual work to do to correct those geometries. Your best bet is to fire up QGIS and get started with digitizing tools to clean things up.

Executing a spatial join and assigning point attributes to a polygon

We'll now get back to some more golf action where we would like to execute a spatial attribute join. We're given a situation where we have a set of polygons, in this case, these are in the form of golf greens without any hole number. Our hole number is stored in a point dataset that is located spatially within the green of each hole. We would like to assign each green its appropriate hole number based on its location within the polygon.

The OSM data from the Pebble Beach Golf Course located in Monterey California is our source data. This golf course is one the great golf courses on the PGA tour and is well mapped in OSM.

If you are interested in getting golf course data yourself from OSM, it is recommended that you use the great Overpass API at `http://overpass-turbo.eu/`. This site enables you to export the OSM data as GeoJSON or KML, for example.

To download all the golf-specific OSM data, you will need to correct tags. To do this, simply copy and paste the following Overpass API query into the query window located on the left hand side, then click on `Download`:

```
/*
    This query looks for nodes, ways, and relations
    using the given key/value combination.
    Choose your region and hit the Run button above!
*/
[out:json][timeout:25];
// gather results
(
    // query part for: "leisure=golf_course"
    node["leisure"="golf_course"]({{bbox}});
    way["leisure"="golf_course"]({{bbox}});
    relation["leisure"="golf_course"]({{bbox}});

    node["golf"="pin"]({{bbox}});
    way["golf"="green"]({{bbox}});
    way["golf"="fairway"]({{bbox}});
    way["golf"="tee"]({{bbox}});
    way["golf"="fairway"]({{bbox}});
    way["golf"="bunker"]({{bbox}});
    way["golf"="rough"]({{bbox}});
    way["golf"="water_hazard"]({{bbox}});
    way["golf"="lateral_water_hazard"]({{bbox}});
    way["golf"="out_of_bounds"]({{bbox}});
    way["golf"="clubhouse"]({{bbox}});
    way["golf"="ground_under_repair"]({{bbox}});

);
// print results
out body;
>;
out skel qt;
```

Getting ready

Importing our data into PostGIS will be the first step to execute our spatial query. This time around, we will use the `shp2pgsql` tool to import our data to change things a little since there are so many ways to get data into PostGIS. The `shp2pgsql` tool is definitely the most well-tested and common way to import Shapefiles into PostGIS. Let's get going and perform this import once again, executing this tool directly from the command line.

For Windows users, this should work, but check that the paths are correct or that `shp2pgsql.exe` is in your system path variable. By doing this, you save having to type the full path to execute.

 I assume that you are running the following command when you are in the `/ch04/code` folder:

```
shp2pgsql -s 4326 ..\geodata\shp\pebble-beach-ply-
greens.shp geodata.pebble_beach_greens | psql -h
localhost -d py_geoan_cb -p 5432 -U pluto
```

On a Linux machine your command is basically the same without the long path, assuming that your system links were all set up when you installed PostGIS in *Chapter 1, Setting Up Your Geospatial Python Environment*.

Next up, we need to import our points with the attributes, so let's get to it as follows:

```
shp2pgsql -s 4326 ..\geodata\shp\pebble-beach-pts-hole-num-green.shp
geodata.pebble_beach_hole_num | psql -h localhost -d py_geoan_cb -p 5432
-U postgres
```

That's that! We now have our points and polygons available in the PostGIS Schema `geodata` setting, which sets the stage for our spatial join.

How to do it...

1. The core work is done once again inside our PostGIS query string, assigning the attributes to the polygons, so follow along:

```python
#!/usr/bin/env python
# -*- coding: utf-8 -*-

import psycopg2

# Database Connection Info
db_host = "localhost"
db_user = "pluto"
db_passwd = "stars"
db_database = "py_geoan_cb"
```

```
    db_port = "5432"

    # connect to DB
    conn = psycopg2.connect(host=db_host, user=db_user, port=db_port,
    password=db_passwd, database=db_database)

    # create a cursor
    cur = conn.cursor()

    # assign polygon attributes from points
    spatial_join = """  UPDATE geodata.pebble_beach_greens AS g
                        SET
                            name = h.name
                        FROM
                            geodata.pebble_beach_hole_num AS h
                        WHERE
                            ST_Contains(g.geom, h.geom);
                    """
cur.execute(spatial_join)
conn.commit()

    # close cursor
    cur.close()

    # close connection
    conn.close()
```

How it works...

The query is straightforward enough; we'll use the UPDATE standard SQL command to update the values in the name field of our table, geodata.pebble_beach_greens, with the hole numbers located in the pebble_beach_hole_num table.

We follow up by setting the name value from our geodata.pebble_beach_hole_num table, where the field name also exists and holds our needed attribute values.

Our WHERE clause uses the PostGIS query, ST_Contains, to return True if the point lies inside our greens, and if so, it will update our values.

This was easy and demonstrates the great power of spatial relationships.

Conducting a complex spatial analysis query using ST_Distance()

Now let's check for a more complex query in PostGIS to get our spatial juices flowing. We want to locate all the golf courses that are either inside or within 5 km of a national park or protected area. Plus, the golf course must be within 2 km of a city. The city data is derived from the tags in OSM where the *tag place = city*.

The national parks and protected areas for this query belong to the Government of Canada. Our golf courses and datasets of cities are derived from an OSM located in British Columbia and Alberta.

Getting ready

We need the data of all the national parks and protected areas in Canada, so go and make sure they're located in the `/ch04/geodata/` folder.

The original data is located at `http://ftp2.cits.rncan.gc.ca/pub/geott/frameworkdata/protected_areas/1M_PROTECTED_AREAS.shp.zip` for download if you do not already have the `/geodata` folder downloaded from GitHub.

The other datasets needed are the cities and golf courses that can be obtained from OSM. These two files are the GeoJSON files located in the /ch04/geodata/ folder and are called `osm-golf-courses-bc-alberta.geojson` and `osm-place-city-bc-alberta.geojson`.

We will now import the downloaded data into our database:

 Ensure that you are currently in the `/ch04/code` folder when you run the following commands; otherwise, adjust the paths as necessary.

1. Starting with the OSM golf courses from British Columbia and Alberta, run this command-line call to ogr2ogr. Windows users need to note that they can either switch the slashes to backslashes or include the full path to GeoJSON:

   ```
   ogr2ogr -f PostgreSQL PG:"host=localhost user=postgres port=5432
   dbname=py_geoan_cb password=air" ../geodata/geojson/osm-golf-
   courses-bc-alberta.geojson -nln geodata.golf_courses_bc_alberta
   ```

2. Now, we'll run the same command again to import the cities:

   ```
   ogr2ogr -f PostgreSQL PG:"host=localhost user=postgres port=5432
   dbname=py_geoan_cb password=air" ../geodata/geojson/osm-place-
   city-bc-alberta.geojson -nln geodata.cities_bc_alberta
   ```

3. Last but not least, we'll need to import the protected areas and national parks of Canada using the `shp2pgsql` command line. Here, note that we need to use the `-W latin1` option to specify the required encoding. The data you acquire is for all of Canada and not just BC and Alberta:

```
shp2pgsql -s 4326 -W latin1 ../geodata/shp/protarea.shp geodata.
parks_pa_canada | psql -h localhost -d py_geoan_cb -p 5432 -U
pluto
```

Now we have all three tables in our database and we can execute our analysis script.

How to do it...

1. Let's see what the code looks like:

```python
#!/usr/bin/env python
# -*- coding: utf-8 -*-

import psycopg2
import json
import pprint
from geojson import loads, Feature, FeatureCollection

# Database Connection Info
db_host = "localhost"
db_user = "pluto"
db_passwd = "stars"
db_database = "py_geoan_cb"
db_port = "5432"

# connect to DB
conn = psycopg2.connect(host=db_host, user=db_user, port=db_port,
password=db_passwd, database=db_database)

# create a cursor
cur = conn.cursor()

complex_query = """
    SELECT
      ST_AsGeoJSON(st_centroid(g.wkb_geometry)) as geom,
c.name AS city, g.name AS golfclub, p.name_en AS park,
    ST_Distance(geography(c.wkb_geometry),
geography(g.wkb_geometry)) AS distance,
    ST_Distance(geography(p.geom),
geography(g.wkb_geometry)) AS distance
      FROM
```

```
        geodata.parks_pa_canada AS p,
        geodata.cities_bc_alberta AS c
        JOIN
        geodata.golf_courses_bc_alberta AS g
        ON
            ST_DWithin(geography(c.wkb_geometry),
geography(g.wkb_geometry),4000)
        WHERE
            ST_DWithin(geography(p.geom),
geography(g.wkb_geometry),5000)
                        """
# WHERE c.population is not null and e.name is not null
# execute the query
cur.execute(complex_query)

# return all the rows, we expect more than one
validity_results = cur.fetchall()

# an empty list to hold each feature of our feature collection
new_geom_collection = []

# loop through each row in result query set and add to my
feature collection
# assign name field to the GeoJSON properties
for each_result in validity_results:
    geom = each_result[0]
    city_name = each_result[1]
    course_name = each_result[2]
    park_name = each_result[3]
    dist_city_to_golf = each_result[4]
    dist_park_to_golf = each_result[5]
    geoj_geom = loads(geom)
    myfeat = Feature(geometry=geoj_geom,
properties={'city': city_name, 'golf_course': course_name,
                        'park_name': park_name, 'dist_to
city': dist_city_to_golf,
                        'dist_to_park':
dist_park_to_golf})
    new_geom_collection.append(myfeat)  # use the geojson
module to create the final Feature Collection of features
created from for loop above

my_geojson = FeatureCollection(new_geom_collection)
```

```
pprint.pprint(my_geojson)

# define the output folder and GeoJSon file name
output_geojson_buf =
"../geodata/golfcourses_analysis.geojson"


# save geojson to a file in our geodata folder
def write_geojson():
    fo = open(output_geojson_buf, "w")
    fo.write(json.dumps(my_geojson))
    fo.close()

# run the write function to actually create the GeoJSON
file
write_geojson()

# close cursor
cur.close()

# close connection
conn.close()
```

How it works...

Let's go step by step through the SQL query:

▸ We'll start with defining what columns we want our query to return and from which tables. Here, we'll define that we want the golf club geometry as a point, city name, golf club name, park name, distance between a city and golf club, and finally, distance between a park and golf club. The geometry that we return is of the golf course as a point, hence `ST_Centroid`, which returns the middle point of the golf course and then outputs this as the GeoJSON geometry.

▸ The `FROM` clause sets our parks and tables of cities and assigns them an alias name with `SQL AS`. We then `JOIN` the golf courses based on the distance using `ST_DWithin()` so that we can locate distances that are less than 4 km between a city and golf course.

▸ The `WHERE` clause, `ST_DWithin()`, enforces the last requirement that the distance between a park and golf course cannot be more than 5 km.

The SQL does all of our heavy lifting to return the correct spatial analysis results. The next step is to use Python to output our results as valid GeoJSON in order to view our newly found golf courses. Each attribute property is then identified by its array location in the query and assigned a name for the GeoJSON output. In the end, we'll output a .geojson file that you can visualize directly in GitHub at `https://github.com/mdiener21/python-geospatial-analysis-cookbook/blob/master/ch04/geodata/golfcourses_analysis.geojson`.

5

Vector Analysis

In this chapter, we will cover the following topics:

- ► Clipping LineStrings to an area of interest
- ► Splitting polygons with lines
- ► Finding the location of a point on a line using linear referencing
- ► Snapping a point to the nearest line
- ► Calculating 3D ground distance and total elevation gain

Introduction

Vector data analysis is used in many, many application areas, starting from measuring the distance from point A to point B all the way through to complex routing algorithms. The first GIS systems were built on vector data and vector analysis, and then later expanded into the raster domain. In this chapter, we will start with simple vector operations, then work our way into a more complex model, chaining the various vector methods together to deliver new data that answers our spatial questions.

This process of data analysis is broken down into a couple of steps starting with an *input* dataset, performing a *spatial operation* on the data such as a buffer analysis, and, finally, we'll have some *output* in the form of a new dataset. The following diagram shows the flow of analysis in the simplest model form:

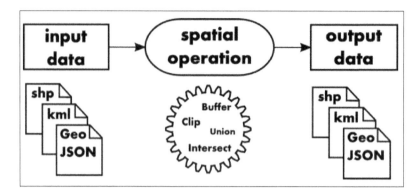

Converting simple questions into spatial operation methods and models takes experience and is not a simple task. For example, you may come across a simple task such as, "Identify and locate how many residential land parcels were affected by the flood." This would translate into the following:

▸ Firstly, an input dataset in the form of a flood polygon that defines the affected floods areas

▸ Secondly, the input dataset represents cadaster polygons

▸ Our spatial operation is an intersection function

▸ All of this results in a new polygon dataset

This would result in a spatial model that could look like this:

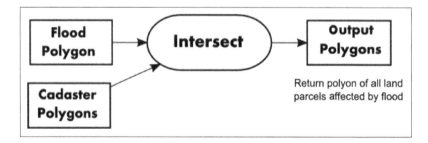

To tackle more complex questions, spatial modeling simply starts chaining more inputs along with more operations that output new data feeding into other new operations. This then leads us to a final set or sets of data.

Clipping LineStrings to an area of interest

A project involving spatial data is typically geographically limited to within a specified boundary area, the so-called *project area*. The input data can come from multiple sources and usually extends outside the project area. Removing this excess data is sometimes critical to speed up spatial processes, and at the same time, it reduces data volume. Reductions in data volumes can also result in secondary speed-ups, for example, less time to transfer or copy the data.

In this recipe, we will take a boundary polygon represented by a circle Shapefile, and then remove all excess LineStrings that are outside this circle.

This process of clipping will remove all lines outside the clip area—that is, our project area of interest.

A standard function called `clip` performs an intersection spatial operation. This is slightly different from a normal intersection function. The clip will NOT or should not retain the attributes attached to the clip area. Clipping involves two input datasets; the first defines the boundary that we want to clip our data to, and the second defines the data that will be clipped. Both sets contain attributes and these attributes of the clipping boundary are usually not included in a clip operation.

The new clipped data will only have the attributes from the original input dataset, excluding all the attributes from the clip polygon.

An `intersection` function will find geometries that overlap and output only lines within a circle. To demonstrate this concept better, the following graphical representation shows what we are going to achieve.

To demonstrate a simple clip operation, we will take a single LineString and polygon defining a clip boundary and perform a quick intersect. The result will look like what's represented in the following screenshot and can be viewed as a live web map in your browser. Refer to the HTML file located at /code/html/ch05-01-clipping.html to check out the result.

When running the simple intersection function, the line is cut into two new LineStrings as in the preceding screenshot.

Our second result will use two Shapefiles that represent our inputs. Our real data from OpenStreetMapconverted is converted to a Shapefile format for our input and output. The circle defines our polygon area of interest, while the road LineStrings are what we want to clip. Our result will be in the form of a new Shapefile that only shows the roads that are inside the circle.

Getting ready

This recipe is in two parts. The first part is a simple clip demonstration using two GeoJSON files consisting of a single LineString and polygon. The second part uses data from OSM and can be found in your `/ch05/geodata` folder containing the circle polygon that represents our area of interest named `clip_area_3857.shp`. The `roads_london_3857.shp` file represents our input Shapefile of lines that we will clip to the circle polygon.

To visualize the first part, we use the leaflet JavaScript library in a very basic HTML page. Our second resulting Shapefile can then be opened with QGIS to see the resulting clipped set of roads.

How to do it...

We have two sets of code examples ahead of us. The first is a simple self-made set of GeoJSON inputs that are clipped and outputted as a GeoJSON representation. This is then visualized using a web page with the help of Leaflet JS.

The second code example takes in two Shapefiles and returns a clipped Shapefile that you can view using QGIS. Both examples use the same method and demonstrate how a clipping function works.

1. Now, let's take a look at the first code example:

```python
#!/usr/bin/env python
# -*- coding: utf-8 -*-

import os
import json
from shapely.geometry import asShape

# define output GeoJSON file
res_line_intersect = os.path.realpath("../geodata/
ch05-01-geojson.js")

# input GeoJSON features
simple_line = {"type":"FeatureCollection","features":[{"type":"Fea
ture","
properties":{"name":"line to
clip"},"geometry":{"type":"LineString","coordinates":
[[5.767822265625,50.14874640066278],[11.901806640625,
50.13466432216696],[4.493408203125,48.821332549646634]]}}]}
```

```
clip_boundary =
{"type":"FeatureCollection","features":[{"type":"Feature",
"properties":{"name":"Clipping boundary circle"},
"geometry":{"type":"Polygon","coordinates":
[[[6.943359374999999,50.45750402042058],[7.734374999999999,
51.12421275782688],[8.96484375,51.316880504045876],
[10.1513671875,51.34433866059924],[10.8544921875,
51.04139389812637],[11.25,50.56928286558243],
[11.25,49.89463439573421],[10.810546875,
49.296471602658094],[9.6240234375,49.03786794532644],
[8.1298828125,49.06666839558117],[7.5146484375,
49.38237278700955],[6.8994140625,49.95121990866206],
[6.943359374999999,50.45750402042058]]]}}]}

# create shapely geometry from FeatureCollection
# access only the geomety part of GeoJSON
shape_line =
asShape(simple_line['features'][0]['geometry'])
shape_circle =
asShape(clip_boundary['features'][0]['geometry'])

# run the intersection
shape_intersect = shape_line.intersection(shape_circle)

# define output GeoJSON dictionary
out_geojson = dict(type='FeatureCollection', features=[])

# generate GeoJSON features
for (index_num, line) in enumerate(shape_intersect):
    feature = dict(type='Feature',
properties=dict(id=index_num))
        feature['geometry'] = line.__geo_interface__
        out_geojson['features'].append(feature)

# write out GeoJSON to JavaScript file
# this file is read in our HTML and
# displayed as GeoJSON on the leaflet map
# called /html/ch05-01-clipping.html
with open(res_line_intersect, 'w') as js_file:
    js_file.write('var big_circle = {0}'.format(json.dumps(clip_
boundary)))
    js_file.write("\n")
    js_file.write('var big_linestring =
{0}'.format(json.dumps(simple_line)))
    js_file.write("\n")
    js_file.write('var simple_intersect =
{0}'.format(json.dumps(out_geojson)))
```

This ends our first code demonstration using a simple self-made GeoJSON LineString that's clipped to a simple polygon. This quick recipe is found in the `/code/ch05-01-1_clipping_simple.py` file. After you run this file, you can go ahead and open the `/code/html/ch05-01-clipping.html` file in your local web browser to see the results.

It works by defining an output JavaScript file that is used to visualize our clipped results. This is followed by our input clipping areas and the LineString to be clipped as GeoJSON. We'll convert our GeoJSON to shapely geometry objects with the `ashape()` function so that we can run the intersection. The resulting intersection geometry is then converted from a shapely geometry into a GeoJSON file that is written to our output JavaScript file, which is used inside the `.html` file for visualization with Leaflet.

2. To begin our second code example located in the `/code/ch05-01-2_clipping.py` file, we will input two Shapefiles, create a new set of roads that are clipped to our circle polygon, and export them as Shapefiles:

```python
#!/usr/bin/env python
# -*- coding: utf-8 -*-

import shapefile
import geojson
import os
# used to import dictionary data to shapely
from shapely.geometry import asShape
from shapely.geometry import mapping

# open roads Shapefile that we want to clip with pyshp
roads_london =
shapefile.Reader(r"../geodata/roads_london_3857.shp")

# open circle polygon with pyshp
clip_area =
shapefile.Reader(r"../geodata/clip_area_3857.shp")

# access the geometry of the clip area circle
clip_feature = clip_area.shape()

# convert pyshp object to shapely
clip_shply = asShape(clip_feature)

# create a list of all roads features and attributes
roads_features = roads_london.shapeRecords()

# variables to hold new geometry
roads_clip_list = []
```

```
roads_shply = []

# run through each geometry, convert to shapely geom and
intersect
for feature in roads_features:
    roads_london_shply =
asShape(feature.shape.__geo_interface__)
    roads_shply.append(roads_london_shply)
    roads_intersect = roads_london_shply.intersection(clip_shply)

    # only export linestrings, shapely also created points
    if roads_intersect.geom_type == "LineString":
        roads_clip_list.append(roads_intersect)

# open writer to write our new shapefile too
pyshp_writer = shapefile.Writer()

# create new field
pyshp_writer.field("name")

# convert our shapely geometry back to pyshp, record for
record
for feature in roads_clip_list:
    geojson = mapping(feature)

    # create empty pyshp shape
    record = shapefile._Shape()

    # shapeType 3 is linestring
    record.shapeType = 3
    record.points = geojson["coordinates"]
    record.parts = [0]

    pyshp_writer._shapes.append(record)
    # add a list of attributes to go along with the shape
    pyshp_writer.record(["empty record"])

# save to disk
pyshp_writer.save(r"../geodata/roads_clipped2.shp")
```

How it works...

For this recipe, we'll use Shapely for our spatial operation and pyshp to read in and out of our Shapefiles.

We'll begin with the import of the road LineStrings and our circle polygon for the demo project area. We'll use the `pyshp` module to handle the Shapefile input/output. `Pyshp` allows us to access the Shapefile bounds, feature geometry, feature attributes, and more.

Our first task is to convert the `pyshp` geometry object into something that `Shapely` can understand. We'll use the `shape()` function to get the `pyshp` geometry followed by the Shapely `asShape()` function. Next, we'll want all the records of roads so that we can use `shapeRecords()` to return these records for us.

Now, we'll get ourselves ready to perform the actual clipping by setting up two list variables to store our new data. The *for* loop runs over each record, that is, each line in the road dataset, converts it to a shapely geometry object using `geo_interface`, and is built in the pyshp function. This is then followed by the actual `intersection` shapely function that only returns geometry that *intersects* our circle. Finally, we'll check to see if the intersection geometry is a LineString. If it is, we'll append it to our output list.

> During an intersection operation, Shapely will return points and LineStrings in a geometry collection. The reason for this is that, if two LineStrings touch at the ends, for example, or overlap each other, it will generate a point intersection location plus any overlapping segments.

At last, we can write out our new dataset to a new Shapefile. Using the pyshp `writer()` function, we create a new object and give it one single field called `name`. Looping through each feature, we can create a GeoJSON object using the shapely mapping function and an empty pyhsp record that we will add to it in a moment. We want to add the point coordinates from our GeoJSON and append them together.

Exiting the loop, we'll save our new Shapefile `roads_clipped.shp` to disk.

Splitting polygons with lines

Typically, in GIS, we work with data that influences other data in some form due to their inherit spatial relationship. This means that we need to work with one dataset to edit, update, and even delete another dataset. A typical example of this is an administrative boundary, which is a polygon that you cannot see on a physical surface but that influences feature information it crosses such as a lake. If we have a lake polygon and an administrative boundary, we might want to know how many square meters of lake belongs to each administrative boundary.

Another example could be a forest polygon that contains one species of trees that crosses a river. We might want to know the area on either side of the river. In the first scenario, we need to transform our administrative boundaries into LineStrings and then perform the cut.

To see what this looks like, take a look at this spoiler on how the results will look up front since we all like a good visual.

Getting ready

For this recipe, we will once again use our GeoJSON LineString and polygon from the previous recipe. These homemade geometries will cut up our polygon into three new polygons. Be sure to fire up your virtual environment with the `workon pygeoan_cb` command.

How to do it...

1. This code example is located at `/code/ch05-02_split_poly_with_line.py` as follows:

    ```
    #!/usr/bin/env python
    # -*- coding: utf-8 -*-
    from shapely.geometry import asShape
    from shapely.ops import polygonize
    import json
    ```

```python
import os

# define output GeoJSON file
output_result = os.path.realpath("../geodata/ch05-02-geojson.js")

# input GeoJSON features
line_geojs =
{"type":"FeatureCollection","features":[{"type":"Feature",
"properties":{"name":"line to clip"},
"geometry":{"type":"LineString","coordinates":
[[5.767822265625,50.14874640066278],
[11.901806640625,50.13466432216696],
[4.493408203125,48.821332549646634]]}}]}
poly_geojs =
{"type":"FeatureCollection","features":
[{"type":"Feature","properties":{"name":"Clipping boundary
circle"},"geometry":{"type":"Polygon","coordinates":[[[6.94
3359374999999,50.45750402042058],[7.734374999999999,
51.12421275782688],[8.96484375,51.316880504045876],
[10.1513671875,51.34433866059924],[10.8544921875,
51.04139389812637],[11.25,50.56928286558243],
[11.25,49.89463439573421],[10.810546875,
49.296471602658094],[9.6240234375,49.03786794532644],
[8.1298828125,49.06666839558117],[7.5146484375,
49.38237278700955],[6.8994140625,49.95121990866206],
[6.943359374999999,50.45750402042058]]]}}]}

# create shapely geometry from FeatureCollection
# access only the geomety part of GeoJSON
cutting_line = asShape(line_geojs['features'][0]['geometry'])
poly_to_split = asShape(poly_geojs['features'][0]['geometry'])

# convert circle polygon to linestring of circle boundary
bndry_as_line = poly_to_split.boundary

# combine new boundary lines with the input set of lines
result_union_lines = bndry_as_line.union(cutting_line)

# re-create polygons from unioned lines
new_polygons = polygonize(result_union_lines)

# stores the final split up polygons
new_cut_ply = []

# identify which new polygon we want to keep
```

```
for poly in new_polygons:
    # check if new poly is inside original otherwise ignore
      it
    if poly.centroid.within(poly_to_split):
        print ("creating polgon")
        # add only polygons that overlap original for
          export
        new_cut_ply.append(poly)
    else:
        print ("This polygon is outside of the input
features")

# define output GeoJSON dictionary
out_geojson = dict(type='FeatureCollection', features=[])

# generate GeoJSON features
for (index_num, geom) in enumerate(new_cut_ply):
    feature = dict(type='Feature',
properties=dict(id=index_num))
    feature['geometry'] = geom.__geo_interface__
    out_geojson['features'].append(feature)

# write out GeoJSON to JavaScript file
# this file is read in our HTML and
# displayed as GeoJSON on the leaflet map
# called /html/ch05-02.html
with open(output_result, 'w') as js_file:
    js_file.write('var cut_poly_result =
{0}'.format(json.dumps(out_geojson)))
```

How it works...

Now the actual splitting of the polygons takes place in our `/ch05/code/ch05-02_split_ poly_with_line.py` script.

The basic methodology to split a polygon based on a LineString follows this simple algorithm. First, we'll take our input polygon and convert the boundaries of this polygon into a new LineString dataset. Next up, we'll combine the LineString we want to use to cut the newly generated polygon LineStrings of boundaries. Finally, we use the `polygonize` method to rebuild polygons based on the new union set of LineStrings.

This rebuilding of polygons results in extra polygons that are created outside the original polygon. To identify these polygons, we'll use a simple trick. We can simply generate a `centroid` point inside each newly created polygon and then check to see if this point is inside the original polygon using the `within` predicate. If the point is not inside the original polygon, the predicate returns `False` and we do not need to include this polygon in our output.

Finding the location of a point on a line using linear referencing

The use of linear referencing is widespread, ranging from storing bus routes to oil and gas pipelines. Our ability to locate any position along a line based on a distance value from the start of the line is done using the interpolation methodology. We want to interpolate a point location at any position along a line. To determine the position, we'll use simple mathematics to calculate the position along a line based on the distance from the starting coordinate.

For our calculation, we'll measure the length of the line and find a coordinate located at a specified length from the start of the line. However, the question of where the start of the line is will soon arise. The starting point of the line is the first coordinate in the LineString's array of vertexes that make up the LineString because a LineString is nothing more than a collection of points chained together.

This will lead nicely to our next recipe, which is a little more complex.

How to do it...

1. This is our shortest code snippet; check it out:

```python
#!/usr/bin/env python
# -*- coding: utf-8 -*-
from shapely.geometry import asShape
import json
import os
from pyproj import Proj, transform


# define the pyproj CRS
# our output CRS
wgs84 = Proj("+init=EPSG:4326")
# output CRS
pseudo_mercator = Proj("+init=EPSG:3857")


def transform_point(in_point, in_crs, out_crs):
    """
    export a Shapely geom to GeoJSON and
    transform to a new coordinate system with pyproj
    :param in_point: shapely geometry as point
    :param in_crs: pyproj crs definition
    :param out_crs:  pyproj output crs definition
    :return: GeoJSON transformed to out_crs
    """
```

```
        geojs_geom = in_point.__geo_interface__

        x1 = geojs_geom['coordinates'][0]
        y1 = geojs_geom['coordinates'][1]

        # transform the coordinate
        x, y = transform(in_crs, out_crs, x1, y1)

        # create output new point
        new_point = dict(type='Feature', properties=dict(id=1))
        new_point['geometry'] = geojs_geom
        new_coord = (x, y)
        # add newly transformed coordinate
        new_point['geometry']['coordinates'] = new_coord

        return new_point


    def transform_linestring(orig_geojs, in_crs, out_crs):
        """
        transform a GeoJSON linestring to
          a new coordinate system
        :param orig_geojs: input GeoJSON
        :param in_crs: original input crs
        :param out_crs: destination crs
        :return: a new GeoJSON
        """
        line_wgs84 = orig_geojs
        wgs84_coords = []
        # transfrom each coordinate
        for x, y in orig_geojs['geometry']['coordinates']:
            x1, y1 = transform(in_crs, out_crs, x, y)
            line_wgs84['geometry']['coordinates'] = x1, y1
            wgs84_coords.append([x1, y1])

        # create new GeoJSON
        new_wgs_geojs = dict(type='Feature', properties={})
        new_wgs_geojs['geometry'] = dict(type='LineString')
        new_wgs_geojs['geometry']['coordinates'] = wgs84_coords

        return new_wgs_geojs


    # define output GeoJSON file
```

```python
output_result = os.path.realpath("../geodata/
ch05-03-geojson.js")

line_geojs = {"type": "Feature", "properties": {},
"geometry": {"type": "LineString", "coordinates":
[[-13643703.800790818,5694252.85913249],
[-13717083.34794459,6325316.964654908]]}}

# create shapely geometry from FeatureCollection
shply_line = asShape(line_geojs['geometry'])

# get the coordinates of each vertex in our line
line_original = list(shply_line.coords)
print line_original

# showing how to reverse a linestring
line_reversed = list(shply_line.coords)[::-1]
print line_reversed

# example of the same reversing function on a string for
example
hello = 'hello world'
reverse_hello = hello[::-1]
print reverse_hello

# locating the point on a line based on distance from line
start
# input in meters = to 360 Km from line start
point_on_line = shply_line.interpolate(360000)

# transform input linestring and new point
# to wgs84 for visualization on web map
wgs_line = transform_linestring(line_geojs,
pseudo_mercator, wgs84)
wgs_point = transform_point(point_on_line, pseudo_mercator,
wgs84)

# write to disk the results
with open(output_result, 'w') as js_file:
    js_file.write('var point_on_line =
{0}'.format(json.dumps(wgs_point)))
    js_file.write('\n')
    js_file.write('var in_linestring =
{0}'.format(json.dumps(wgs_line)))
```

After executing the `/code/ch05-03_point_on_line.py` file, you should see the following screenshot when you open the `/code/html/ch05-03.html` file in your web browser:

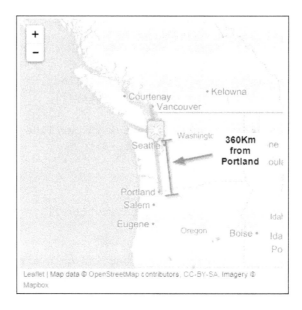

If you would like to reverse the LineString starting and ending points, you can use the `list(shply_line.coords)[::-1]` code to reverse the coordinate order as shown in the preceding code.

How it works...

It all boils down to executing one single line of code to locate a point on a line specified at a certain distance. The shapely interpolate function does this for us. All you need is the shapely LineString geometry and a distance value. The distance value is the distance from the 0,0 start coordinate of the LineString.

Be careful in case the LineString direction is not the correct form in which you want to measure it. This would mean that you need to switch the LineString direction. Take a look at the `line_reversed` variable that holds the original LineString with a reversed order. To do the `reverse`, we'll use a simple python string operation, `[::-1]`, to reverse our LineString list.

You can see this in action with our print statement reversing the LineString order on screen as follows:

```
[(-13643703.800790818, 5694252.85913249), (-13717083.34794459,
6325316.964654908)]
```

```
[(-13717083.34794459, 6325316.964654908), (-13643703.800790818,
5694252.85913249)]
```

See also

If you are interested in more information regarding linear referencing, ESRI has a great reference of use cases and examples at `http://resources.arcgis.com/en/help/ma in/10.1/0039/003900000001000000.htm` and `http://en.wikipedia.org/wiki/ Linear_referencin`.

Snapping a point to the nearest line

Building on our newly gained wisdom from the last recipe, we will now attack another common spatial problem. This super common spatial task is for all the GPS junkies who want their GPS coordinates to snap to an existing road. Imagine that you have some GPS tracks and you want to have these coordinates snap to your base road dataset. To accomplish such a task, we need to snap a point (GPS coordinates) to a line (roads).

The `geos` library is what `Shapely` is built on and can handle this problem with ease. We will combine the use of the `shapely.interpolate` and `shapely.project` functions to snap our point to the true nearest point on the line using linear referencing.

As you can see in the following diagram, our input point is located on the sun icon. The green line is what we want to snap our point to at the nearest location. The gray icon with a point on it is our result that represents the nearest point on the line from our original x position.

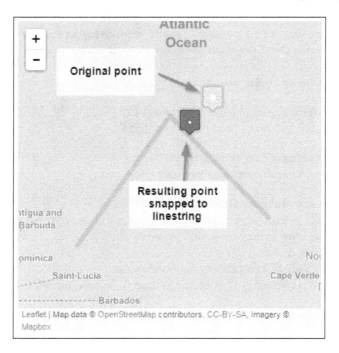

How to do it...

1. Shapely is well suited for snapping a point to the nearest line, so let's get started:

```python
#!/usr/bin/env python
# -*- coding: utf-8 -*-

from shapely.geometry import asShape
import json
import os
from pyproj import Proj, transform

# define the pyproj CRS
# our output CRS
wgs84 = Proj("+init=EPSG:4326")
# output CRS
pseudo_mercator = Proj("+init=EPSG:3857")


def transform_point(in_point, in_crs, out_crs):
    """
    export a Shapely geom to GeoJSON Feature and
    transform to a new coordinate system with pyproj
    :param in_point: shapely geometry as point
    :param in_crs: pyproj crs definition
    :param out_crs: pyproj output crs definition
    :return: GeoJSON transformed to out_crs
    """
    geojs_geom = in_point.__geo_interface__

    x1 = geojs_geom['coordinates'][0]
    y1 = geojs_geom['coordinates'][1]

    # transform the coordinate
    x, y = transform(in_crs, out_crs, x1, y1)

    # create output new point
    out_pt = dict(type='Feature', properties=dict(id=1))
    out_pt['geometry'] = geojs_geom
    new_coord = (x, y)
    # add newly transformed coordinate
    out_pt['geometry']['coordinates'] = new_coord

    return out_pt
```

```python
def transform_geom(orig_geojs, in_crs, out_crs):
    """
    transform a GeoJSON linestring or Point to
      a new coordinate system
    :param orig_geojs: input GeoJSON
    :param in_crs: original input crs
    :param out_crs: destination crs
    :return: a new GeoJSON
    """

    wgs84_coords = []
    # transfrom each coordinate
    if orig_geojs['geometry']['type'] == "LineString":
        for x, y in orig_geojs['geometry']['coordinates']:
            x1, y1 = transform(in_crs, out_crs, x, y)
            orig_geojs['geometry']['coordinates'] = x1, y1
            wgs84_coords.append([x1, y1])
        # create new GeoJSON
        new_wgs_geojs = dict(type='Feature', properties={})
        new_wgs_geojs['geometry'] = dict(type='LineString')
        new_wgs_geojs['geometry']['coordinates'] =
wgs84_coords

        return new_wgs_geojs

    elif orig_geojs['geometry']['type'] == "Point":

        x = orig_geojs['geometry']['coordinates'][0]
        y = orig_geojs['geometry']['coordinates'][1]
        x1, y1 = transform(in_crs, out_crs, x, y)
        orig_geojs['geometry']['coordinates'] = x1, y1
        coord = x1, y1
        wgs84_coords.append(coord)

        new_wgs_geojs = dict(type='Feature', properties={})
        new_wgs_geojs['geometry'] = dict(type='Point')
        new_wgs_geojs['geometry']['coordinates'] =
wgs84_coords

        return new_wgs_geojs
    else:
        print("sorry this geometry type is not supported")
```

```
# define output GeoJSON file
output_result = os.path.realpath("../geodata/ch05-04-
geojson.js")

line =
{"type":"Feature","properties":{},"geometry":{"type":
"LineString","coordinates":
[[-49.21875,19.145168196205297],
[-38.49609375,32.24997445586331],
[-27.0703125,22.105998799750576]]}}
point =
{"type":"Feature","properties":{},"geometry":{"type":
"Point","coordinates":[-33.57421875,32.54681317351514]}}

new_line = transform_geom(line, wgs84, pseudo_mercator)
new_point = transform_geom(point, wgs84, pseudo_mercator)


shply_line = asShape(new_line['geometry'])
shply_point = asShape(new_point['geometry'])

# perform interpolation and project point to line
pt_interpolate = shply_line.interpolate(shply_line.project(shply_
point))

# print coordinates and distance to console
print ("origin point coordinate")
print (point)

print ("interpolted point location")
print (pt_interpolate)

print "distance from origin to interploate point"
print (shply_point.distance(pt_interpolate))

# convert new point to wgs84 GeoJSON
snapped_pt = transform_point(pt_interpolate,
pseudo_mercator, wgs84)

# our original line and point are transformed
# so here they are again in original coords
# to plot on our map
```

```
line_orig =
{"type":"Feature","properties":{},"geometry":{"type":
"LineString","coordinates":
[[-49.21875,19.145168196205297],
[-38.49609375,32.24997445586331],[-
27.0703125,22.105998799750576]]}}
point_orig =
{"type":"Feature","properties":{},"geometry":{"type":
"Point","coordinates":[-33.57421875,32.54681317351514]}}

# write to disk the results
with open(output_result, 'w') as js_file:
    js_file.write('var input_pt =
{0}'.format(json.dumps(snapped_pt)))
    js_file.write('\n')
    js_file.write('var orig_pt =
{0}'.format(json.dumps(point_orig)))
    js_file.write('\n')
    js_file.write('var line =
{0}'.format(json.dumps(line_orig)))
```

How it works...

We'll use a tried and tested methodology called **linear referencing** to do the work. Let's kick it off with the imports needed to do this, including `shapely.geometry asShape`, `json`, and `pyproj`' Pyproj is used to quickly transform our coordinates to and from EPSG: 4326 and EPSG 3857. Shapely works on planar coordinates and cannot work directly with `lat/lon` values.

Extending our functions from the last recipe, we have the `transform_point()` function alongside the `transform_geom()` function. The `transform_point()` function converts a Shapely geometry to GeoJSON and transforms the point coordinate, while the `transform_geom()` function takes GeoJSON in and transforms it to the new coordinate system. Both functions use pyproj to execute the transformations.

Next, we'll define our output GeoJSON file and the input line and point features. Then, we'll execute our two new transform functions followed closely with the conversion into a Shapely geometry object. This new Shapely geometry is then run through the interpolate function.

Interpolate alone does not answer our question. We need to combine its usage with the Shapely `project` function that takes in the original point and projects it onto the line.

We then print out our results to screen and create a new JavaScript file called `/geodata/ch05-04-geojson.js`, used in our `/code/html/ch05-04.html` for viewing. Go ahead and open the HTML file in your browser to see the results.

Take a look at your console to see the print to console statements that show us the point coordinates and distances from the original as follows:

```
>>> python python-geospatial-analysis-cookbook/ch05/code/ch05-04_snap_point2line.py
```

Calculating 3D ground distance and total elevation gain

We've finished finding points on lines and returning points on a line, so now, it is time to calculate the true ground 3D distance that we actually ran or biked along a real 3D road. It is also possible to calculate the elevation profile and we will see this in the *Chapter 7, Raster Analysis*.

Calculating the ground distance sounds easy, but 3D calculations are more complicated to calculate than 2D. Our 3D LineString has a z-coordinate for each vertex that makes up our LineString. Therefore, we need to calculate the 3D distance between each set of coordinates, —that is, from vertex to vertex in our input LineString.

The mathematics to calculate the distance between two 3D Cartesian coordinates is relatively simple and uses the 3D form of the Pythagoras formula:

$$3d\_distance = square\ root\ \sqrt{\ ((x2 - x1)^2 + (y2 - y1)^2 + (z2 - z1)^2)}$$

Here it is in Python:

```
import math
3d_dist = math.sqrt((x2 - x1)**2 + (y2 - y1)**2 + (z2 - z1)**2 )
```

Getting ready

First up, we will get our hands on some 3D data to play with and what better to analyze than the Stage 16 Carcassonne/Bagnères-de-Luchon of the Tour de France 2014 mountain stage, a real killer. Here are some stats from `www.letour.com`, including the 237.5 km length, Michael Rogers' winning time of 6:07:10, the average speed of 38.811 km/h, and the highest point of 1753 m. You will find the data in your folder at `/ch05/geodata/velowire_stage_16_27563_utf8.geojson`.

The original KML was generously provided by Thomas Vergouwen (www.velowire.com) and it is free for us to use with his permission; thanks, Thomas. The original data is located at /ch05/geodata/velowire_stage_16-Carcassonne-Bagneres-de-Luchon.kml. The conversion to *GeoJSON* and the transformation to EPSG:27563 was done using the QGIS save as function.

Now, according to the LA Times web page (http://www.latimes.com/la-sp-g-tour-de-france-stage-elevation-profile-20140722-htmlstory.html), they've quoted a 3895 m elevation gain. As compared to team Strava (http://blog.strava.com/tour-de-france-2014/) where they've stated a 4715 m elevation gain. Now, who is correct and is this 237.5 km ground distance in 3D? Let's find out!

This is the official profile of Stage 16 for your visual pleasure:

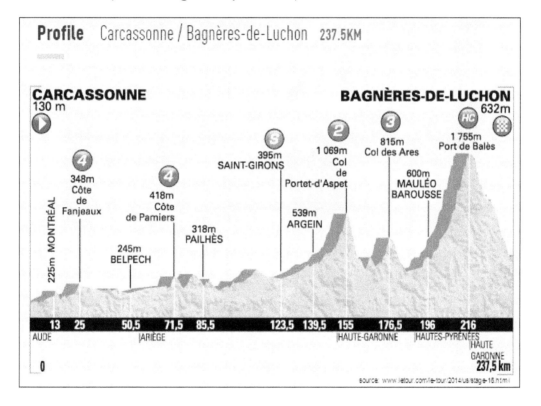

To give you an idea of what accurate and simplified data looks like, take a look at this comparison of the velowire's site's (`www.velowire.com`) KML marked in purple (accurate) and the bikemap site's progression highlighted by yellow line (simplified). If you sum up the differences, the length and elevation are both significantly different for both. For a race that's 237.5 km long, every meter counts when you're planning and attacking on the course. In the following screenshot, you can see the comparison of the velowire site's KML marked in purple and the bikemap site's progression highlighted by the yellow line:

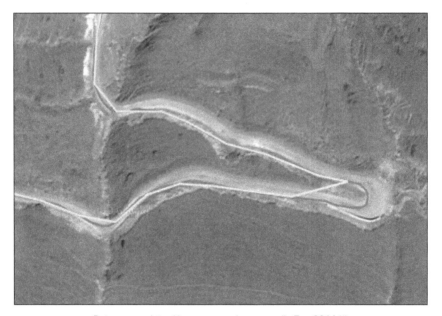

Data source: http://www.mapcycle.com.au/LeTour2014/#

How to do it...

We'll start with looping through each vertex and calculating the 3D distance from one vertex to another in our LineString. Each vertex is nothing more than a point with *x*, *y*, and *z* (3D Cartesian) values.

1. Here is the code to calculate each vertex:

```python
#!/usr/bin/env python
# -*- coding: utf-8 -*-
import math
import os
from shapely.geometry import shape, Point
import json

def pairs(lst):
    """
```

```
    yield iterator of two coordinates of linestring
    :param lst: list object
    :return: yield iterator of two coordinates
    """
    for i in range(1, len(lst)):
        yield lst[i - 1], lst[i]


def calc_3d_distance_2pts(x1, y1, z1, x2, y2, z2):
    """
    :input two point coordinates (x1,y1,z1),(x2,y2,2)
    :param x1: x coordinate first segment
    :param y1: y coordiante first segment
    :param z1: z height value first coordinate
    :param x2: x coordinate second segment
    :param y2: y coordinate second segment
    :param z2: z height value second coordinate
    :return: 3D distance between two input 3D coordinates
    """
    d = math.sqrt((x2 - x1) ** 2 + (y2 - y1) ** 2 + (z2 -
z1) ** 2)
    return d


def readin_json(jsonfile):
    """
    input: geojson or json file
    """
    with open(jsonfile) as json_data:
        d = json.load(json_data)
        return d


geoj_27563_file = os.path.realpath("../geodata/velowire_
stage_16_27563_utf8.geojson")
print (geoj_27563_file)
# create python dict type from geojson file object
json_load = readin_json(geoj_27563_file)

# set start lengths
length_3d = 0.0
length_2d = 0.0

# go through each geometry in our linestring
for f in json_load['features']:
    # create shapely shape from geojson
```

```
        s = shape(f['geometry'])

        # calculate 2D total length
        length_2d = s.length

        # set start elevation
        elevation_gain = 0

        # go through each coordinate pair
        for vert_start, vert_end in pairs(s.coords):
            line_start = Point(vert_start)
            line_end = Point(vert_end)

            # create input coordinates
            x1 = line_start.coords[0][0]
            y1 = line_start.coords[0][1]
            z1 = line_start.coords[0][2]
            x2 = line_end.coords[0][0]
            y2 = line_end.coords[0][1]
            z2 = line_end.coords[0][2]

            # calculate 3d distance
            distance = calc_3d_distance_2pts(x1, y1, z1, x2,
y2, z2)

            # sum distances from vertex to vertex
            length_3d += distance

            # calculate total elevation gain
            if z1 > z2:
                elevation_gain = ((z1 - z2) + elevation_gain )
                z2 = z1
            else:
                elevation_gain = elevation_gain  # no height
change
                z2 = z1


    print ("total elevation gain is: {gain} meters".
    format(gain=str(elevation_gain)))

    # print coord_pair
    distance_3d = str(length_3d / 1000)
    distance_2d = str(length_2d / 1000)
    dist_diff = str(length_3d - length_2d)
```

```
print ("3D line distance is: {dist3d}
meters".format(dist3d=distance_3d))
print ("2D line distance is: {dist2d}
meters".format(dist2d=distance_2d))
print ("3D-2D length difference: {diff}
meters".format(diff=dist_diff))
```

How it works...

We need to transform our original KML file stored in EPSG: 4326 to a planar coordinate system to facilitate our calculations (refer to the upcoming table). So, we'll begin by transforming the KML into EPSG: 27563 NTF Paris / Lambert Sud France. For further information on this, refer to `http://epsg.io/27563`.

To begin with, we'll define three functions for our calculations starting with the `pairs()` function that takes a list and then uses the Python yield generator function to yield two sets of values. The first set of values is the starting x, y, and z coordinates, and the second set includes the ending x, y, and z coordinates of the coordinate pairs that we want to measure.

The `calc_3d_distancte_2pts()` function takes the two coordinate pairs, including the important z value, and calculates the distance between two points in 3D space using the Pythagorean theorem.

Our `readin_json()` function inputs a path to a file, and in this case, we can point it to our GeoJSON file stored in the `/ch05/geodata` folder. This will return a Python dictionary object for us to work with within the next few steps.

Now, let's define the variables to hold our GeoJSON file, load this file, and set the starting 3D/2D lengths to zero for initialization.

Next up, let's iterate through the GeoJSON LineString features and convert them into a Shapely object so that we can use Shapely to tell us the inherent 2D length as used by our `length_2d` variable and read the coordinates. This is followed by our `for` loop where all the action occurs.

Looping over our new list created by our `pairs()` function, we can loop over each vertex of a LineString. We define the `line_start` and `line_end` variables to identify the start of each new line segment that we need to access with a single LineString feature. We'll follow up by then defining our input parameters to do the 3D distance calculations by parsing our list object with standard Python positional slicing. At last, we'll call the `calc_3d_distance_2pts()` function to give us our distance in 3D.

We need to iteratively sum the distances together from one segment to the next. We can do this by adding the distance to our `length_3d` with += operator. Now, our `length_3d` variable is updated for each segment in the loop, giving us our desired 3D length.

The remaining part of the loop calculates our elevation gain. Our $z1$ and $z2$ elevation values need to be constantly compared to additively add the total elevation gain only if the next value is greater than the last. If not, set them to equal each other and continue to the next z value. The `elevation_gain` variable is then constantly updated to itself if there's no change; otherwise not, the difference between the two elevations is added.

At last, we'll print out our results to the screen; they should look like this:

```
total elevation gain is: 4322.0 meters
3D line distance is: 244.119162551
2D line distance is: 243.55802081
3D-2D length difference: 561.141741137 meters
```

With our data transformed and converted to GeoJSON, the 2D length according to our script is 243.558 km from the velowire KML as compared to 237.5 km from the official race page, which is a difference of 6.058 km. The original KML in EPSG:4326 was 302.805 km long, a difference of over 65 km, hence the necessary transformation. For a better comparison, take a look at this table:

Source + EPSG	2D length	3D length	Difference
Velowire EPSG:4326	302.805 km	This is not calculated	
Velowire EPSG:27563	**243.558 km**	**244.12**	**561.14 m**
Mapcycle EPSG:4326	293.473 km	This data is not available	
Mapcylce EPSG:27563	236.216 km	This data is not available	
Letour official	237.500 km (approximate)	237.500 km (approximate)	

The elevation gain is also very different between different sources.

Source	Elevation gain
Strava (`http://blog.strava.com/`)	4715 m
Los Angeles Times	3895 m
TrainingPeaks (`www.trainingpeaks.com`)	3243 m
The Velowire KML data analysis	4322 m

There's more...

The accuracy of all these calculations is based on the original KML data source. Each data source is/was derived by different people and, possibly, different methods. The more you know about your data source, the more you know about its accuracy. In this case, I assume that the Velowire data source was digitized by hand using Google Earth. Thus, the accuracy is only as accurate as that of the underlying Google Earth imagery and coordinate system, which is EPSG:3857.

6
Overlay Analysis

In this chapter, we will cover the following topics:

▸ Punching holes in polygons with a symmetric difference operation

▸ Union polygons without merging

▸ Union polygons with merging (dissolving)

▸ Performing an identity function (difference + intersection)

Introduction

Discovering how two datasets spatially relate to each other when they are placed over one another is called overlay analysis. An overlay can be compared to a sheet of tracing paper. For example, you could overlay the tracing paper on top of your base map and see what areas overlap each other. This process is and was a game changer in spatial analysis and modeling. Computer-aided GIS computations can therefor automatically identify where two geometry sets spatially touch for example.

The goal of this chapter is to give you a feel for the most common overlay analysis functions, such as unions, intersects, and symmetrical differences. These are based on the **Dimensionally Extended nine intersection model** (**DE-9IM**), which can be found at `http://en.wikipedia.org/wiki/DE-9IM`, and describes our list of possible overlays. All processes that we use or name here are derived using a combination of these nine predicates.

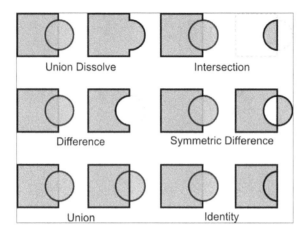

We will explore these topology rules in depth in *Chapter 9, Topology Checking and Data Validation*.

Punching holes in polygons with a symmetric difference operation

Why, oh why would we want to punch holes in polygons and create a donut? Well, this is done for several reasons, for example, you may want to remove a lake polygon from a forest polygon that it overlaps since it sits in the middle of the forest and is, therefore, included in your area calculations.

Another example is where we have a set of polygons representing a golf course's fairways and a second set of polygons representing the greens that overlap these fairways. Our task is to calculate the correct number of square meters of fairways. The greens will create our donuts in a fairway's polygons.

This is translated into spatial operation terminology and means that we need to perform a `symmetric difference` operation or, in ESRI terminology, an "erase" operation.

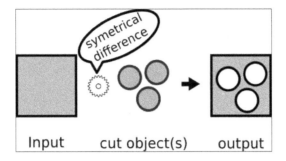

Getting ready

In this example, we will create two sets of visualizations to see our results. Our output will generate **Well Known Text** (**WKT**) that is displayed in your browser using the **Openlayers 3** web mapping client.

For this example, make sure you have all your code downloaded in your /ch06 folder provided at GitHub and have this folder structure containing these files:

```
code
|   ch06-01_sym_diff.py
|   foldertree.txt
|   utils.py
|
+---ol3
    +---build
    |       ol-debug.js
    |       ol-deps.js
    |       ol.js
    |
    +---css
    |       layout.css
    |       ol.css
    |
    +---data
    |       my_polys.js
    |
    +---html
    |       ch06-01_sym_diff.html
    |
    +---js
```

```
    |          map_sym_diff.js
    |
    +---resources
    |       jquery.min.js
    |       logo-32x32-optimized.png
    |       logo-32x32.png
    |       logo.png
    |       textured_paper.jpeg
    |
    +---bootstrap
        +---css
        |           bootstrap-responsive.css
        |           bootstrap-responsive.min.css
        |           bootstrap.css
        |           bootstrap.min.css
        |
        +---img
        |           glyphicons-halflings-white.png
        |           glyphicons-halflings.png
        |
        +---js
                    bootstrap.js
                    bootstrap.min.js

geodata
    pebble-beach-fairways-3857.geojson
    pebble-beach-greens-3857.geojson
    results_sym_diff.js
```

With the folder structure in place, when you run the code, all inputs and outputs will find their correct home.

How to do it...

We want to run this code from the command line as usual, which runs in your virtual environment:

1. Execute the following statement from your /ch06/code folder:

   ```
   >> python Ch06-01_sym_diff.py
   ```

2. The following code is where interesting operations take place with Shapely:

   ```
   #!/usr/bin/env python
   # -*- coding: utf-8 -*-
   ```

```
import json
from os.path import realpath
from shapely.geometry import MultiPolygon
from shapely.geometry import asShape
from shapely.wkt import dumps


# define our files input and output locations
input_fairways = realpath("../geodata/
pebble-beach-fairways-3857.geojson")
input_greens = realpath("../geodata/
pebble-beach-greens-3857.geojson")
output_wkt_sym_diff =
realpath("ol3/data/results_sym_diff.js")


# open and load our geojson files as python dictionary
with open(input_fairways) as fairways:
    fairways_data = json.load(fairways)

with open(input_greens) as greens:
    greens_data = json.load(greens)

# create storage list for our new shapely objects
fairways_multiply = []
green_multply = []

# create shapely geometry objects for fairways
for feature in fairways_data['features']:
    shape = asShape(feature['geometry'])
    fairways_multiply.append(shape)

# create shapely geometry objects for greens
for green in greens_data['features']:
    green_shape = asShape(green['geometry'])
    green_multply.append(green_shape)

# create shapely MultiPolygon objects for input analysis
fairway_plys = MultiPolygon(fairways_multiply)
greens_plys = MultiPolygon(green_multply)

# run the symmetric difference function creating a new
Multipolygon
result = fairway_plys.symmetric_difference(greens_plys)
```

```
# write the results out to well known text (wkt) with
shapely dump
def write_wkt(filepath, features):
    with open(filepath, "w") as f:
        # create a js variable called ply_data used in html
        # Shapely dumps geometry out to WKT
        f.write("var ply_data = '" + dumps(features) + "'")


# write to our output js file the new polygon as wkt
write_wkt(output_wkt_sym_diff, result)
```

Your output will be available in the `/ch06/code/ol3/html/` folder with the `ch06-01_sym_diff.html` filename. Simply open this file in your local web browser, such as Chrome, Firefox, or Safari. Our output web map was created by modifying the Openlayers 3 example code pages according to our needs. The resulting web map should display the following map in your local web browser:

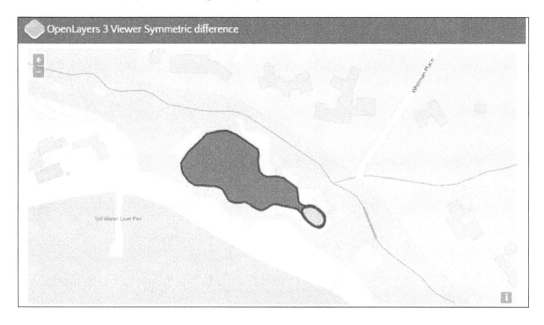

You can now clearly see a hole inside our fairway.

How it works...

To begin with, we use two **GeoJSON** datasets as our input, both with EPSG: 3857 and stemming from the OSM EPSG: 4326. The transformation process is not covered here; take a look at *Chapter 2, Working with Projections*, for further information on how to transform data between two coordinate systems.

Our first task is to read in both the GeoJSON files into Python dictionaries objects using the standard Python `json` module. Next, we set up some empty lists that will store the Shapely geometry objects as a list used for our input to generate the needed `MultiPolygons` for our analysis. We use the Shapely built-in `asShape()` function to create the Shapely geometry objects so that we can perform the spatial operations. This is accomplished by accessing the dictionaries' `['geometry']` element. We then append each geometry to our empty list. This list is then inputted into the Shapely `MultiPolygon()` function that will create a MultiPolygon for us and is used as our inputs.

The actual process of running our `symmetric_difference` happens when we input the `fairways_plys` MultiPolygon as input and the parameter passed is the `greens_ply` MultiPolygon. The output is stored in the `result` variable, which itself is also a MultiPolygon. Not to forget, a MultiPolygon is just a list of polygons that we can iterate over.

Next up, we'll take a look at a function called `write_wkt(filepath, features)`. This outputs our resulting MultiPolygon Shapely geometry to the `Well Known Text (WKT)` format. We do not simply output this `WKT` but instead, create a new JavaScript file, `ol3/data/ch06-01_results_sym_diff.js`, containing our `WKT` output. The code outputs a string that creates a JavaScript variable called `ply_data`. This `ply_data` variable is then used in our HTML file located at `/ch06/code/ol3/html/sym_diff.html` to draw our `WKT` vector layer using Openlayers 3. We then call our function and it executes the write to the `WKT` JavaScript file.

This example is the first that visualizes our results as a web map. In *Chapter 11, Web Analysis with GeoDjango*, we will explore a fully functional web mapping application; for those of you who cannot wait, you may want to jump ahead. Further examples will continue to use Openlayers 3 as our data viewer, moving away from using Matplotlib.

In the end, our simple one-line symmetric difference execution needed a lot of helper code to deal with importing GeoJSON data and exporting the results in a format that could display a web map with Openlayers 3.

Union polygons without merging

To demonstrate what merging is all about, we will take an example from the **National Oceanic and Atmospheric Administration** (**NOAA**) weather data. It provides an awesome minute-by-minute update of Shapefiles for your desire to download data. We will look at a one-week collection of weather warnings, and combine these with state boundaries to see where exactly warnings occurred within a state boundary.

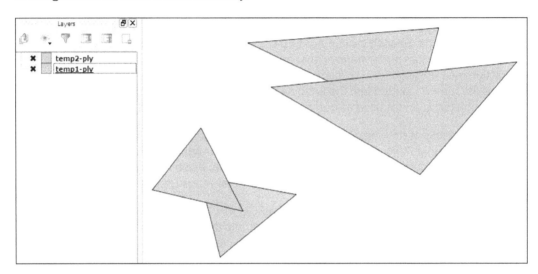

The preceding screenshot shows us the polygons before the union operation in QGIS.

Getting ready

Make sure your virtual environment is, as always, fired up and run the following command:

```
$ source venvs/pygeo_analysis_cookbook/bin/activate
```

Next, switch to your /ch06/code/ folder to find finished code examples or create your empty file in the /ch06/working folder and follow along with the code.

How to do it...

The pyshp and shapely libraries are our two workhorses for this exercise:

1. You can simply run this file in the command prompt to see the results as follows:

   ```
   >> python ch06-02_union.py
   ```

Results can then be opened in the /ch06/code/ol3/html/ch06-02_union. html folder with a double-click to start them in your local web browser. You should see the following web map if everything's gone smoothly:

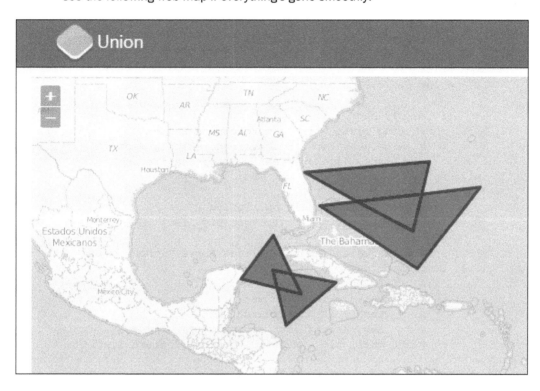

2. Now let's take a look at the code that makes it all happen:

```python
#!/usr/bin/env python
# -*- coding: utf-8 -*-
import json
from os.path import realpath
import shapefile  # pyshp
from geojson import Feature, FeatureCollection
from shapely.geometry import asShape, MultiPolygon
from shapely.ops import polygonize
from shapely.wkt import dumps


def create_shapes(shapefile_path):
    """
    Convert Shapefile Geometry to Shapely MultiPolygon
    :param shapefile_path: path to a shapefile on disk
    :return: shapely MultiPolygon
```

```
        """
        in_ply = shapefile.Reader(shapefile_path)

        # using pyshp reading geometry
        ply_shp = in_ply.shapes()
        ply_records = in_ply.records()
        ply_fields = in_ply.fields
        print ply_records
        print ply_fields

        if len(ply_shp) > 1:
            # using python list comprehension syntax
            # shapely asShape to convert to shapely geom
            ply_list = [asShape(feature) for feature in
ply_shp]

            # create new shapely multipolygon
            out_multi_ply = MultiPolygon(ply_list)

            # # equivalent to the 2 lines above without using
list comprehension
            # new_feature_list = []
            # for feature in features:
            #     temp = asShape(feature)
            #     new_feature_list.append(temp)
            # out_multi_ply = MultiPolygon(new_feature_list)

            print "converting to MultiPolygon: " +
str(out_multi_ply)
        else:
            print "one or no features found"
            shply_ply = asShape(ply_shp)
            out_multi_ply = MultiPolygon(shply_ply)

        return out_multi_ply


def create_union(in_ply1, in_ply2, result_geojson):
    """
    Create union polygon
    :param in_ply1: first input shapely polygon
    :param in_ply2: second input shapely polygon
    :param result_geojson: output geojson file including
full file path
    :return: shapely MultiPolygon
    """
    # union the polygon outer linestrings together
```

```python
        outer_bndry = in_ply1.boundary.union(in_ply2.boundary)

        # rebuild linestrings into polygons
        output_poly_list = polygonize(outer_bndry)

        out_geojson = dict(type='FeatureCollection', features=[])

        # generate geojson file output
        for (index_num, ply) in enumerate(output_poly_list):
            feature = dict(type='Feature',
    properties=dict(id=index_num))
            feature['geometry'] = ply.__geo_interface__
            out_geojson['features'].append(feature)

        # create geojson file on disk
        json.dump(out_geojson, open(result_geojson, 'w'))

        # create shapely MultiPolygon
        ply_list = []
        for fp in polygonize(outer_bndry):
            ply_list.append(fp)

        out_multi_ply = MultiPolygon(ply_list)

        return out_multi_ply


def write_wkt(filepath, features):
    """

    :param filepath: output path for new JavaScript file
    :param features: shapely geometry features
    :return:
    """
    with open(filepath, "w") as f:
        # create a JavaScript variable called ply_data used in
html
        # Shapely dumps geometry out to WKT
        f.write("var ply_data = '" + dumps(features) + "'")


def output_geojson_fc(shply_features, outpath):
    """
    Create valid GeoJSON python dictionary
    :param shply_features: shapely geometries
    :param outpath:
    :return: GeoJSON FeatureCollection File
```

```
        """

    new_geojson = []
    for feature in shply_features:
        feature_geom_geojson = feature.__geo_interface__
        myfeat = Feature(geometry=feature_geom_geojson,
                         properties={'name': "mojo"})
        new_geojson.append(myfeat)

    out_feat_collect = FeatureCollection(new_geojson)

    with open(outpath, "w") as f:
        f.write(json.dumps(out_feat_collect))


if __name__ == "__main__":

    # define our inputs
    shp1 = realpath("../geodata/temp1-ply.shp")
    shp2 = realpath("../geodata/temp2-ply.shp")

    # define outputs
    out_geojson_file =
realpath("../geodata/res_union.geojson")
    output_union = realpath("../geodata/output_union.geojson")
    out_wkt_js = realpath("ol3/data/results_union.js")

    # create our shapely multipolygons for geoprocessing
    in_ply_1_shape = create_shapes(shp1)
    in_ply_2_shape = create_shapes(shp2)

    # run generate union function
    result_union = create_union(in_ply_1_shape,
in_ply_2_shape, out_geojson_file)

    # write to our output js file the new polygon as wkt
    write_wkt(out_wkt_js, result_union)

    # write the results out to well known text (wkt) with
shapely dump
    geojson_fc = output_geojson_fc(result_union,
output_union)
```

How it works...

A quick high-level run-through of what is going on here at the beginning should help clear the air. We have four functions and nine variables within our Python code to split the load of input and output data. The running of our code takes place in the `if __name__ == "main":` call that is found at the end of the code. We start defining two variables to deal with our inputs that we are going to **union** together. These two are our input Shapefiles and the other three outputs are GeoJSON and JavaScript files.

The `create_shapes()` function converts our Shapefile into Shapely `MultiPolygon` geometry objects. Inside the Python class, the list comprehension is used to generate a new list of polygon objects, which are the input list of polygons used to create our output `MultiPolygon`. Next, we'll simply run this function passing in our input Shapefiles.

Our `create_union()` function is up next where we do the real union work. We begin by unioning the two geometry boundaries together that produces a union set of LineStrings and represents the outer bounds of our input polygons. The reason for this is that we do not want to lose the geometries of both polygons, which will, by default, dissolve into one big polygon when simply passed into the Shapely union function. Therefore, we need to rebuild the polygons with the `polygonize()` Shapely function.

The `polygonize` function creates a Python **generator** object, not a simple geometry. This is an *iterator* that's similar to a *list* that we need to loop over to get at the individual polygons it's created for us.

We do exactly this in the next code segment using the `enumerate()` Python function that automatically creates an ID for us for each feature that we use as the id field in the attribute results. After our loop, we use the standard Python `json.dump()` method to export our newly created GeoJSON file and write it to disk using the Python `open()` method in the write mode.

Lastly, in our `create_union()` function, we prepare to output our resulting **union** polygon as a Shapely MultiPolygon object. This is accomplished simply by looping through the `polygonize()` iterator and outputting a list that feeds into the Shapely `MultiPolygon()` function. Finally, we execute the union function, passing in our two input geometries and specifying the output GeoJSON file.

So, we can view our results in our web map as we did in the previous exercise using a small function called `write_wkt()`. This little function takes the file path to the output JavaScript file that we want to create and the MultiPolygon result's geometry. Shapely then dumps the geometry into the Well Known Text format as we write it out to the JavaScript file.

In the end, a small function called `output_geojson_fc()` is used to output another GeoJSON file, this time using the Python `geojson` library. This simply shows you another way to recreate a GeoJSON file. Since GeoJSON is a plain text file, it is possible to create it in many unique ways depending on your personal programming preference.

Union polygons with merging (dissolving)

To demonstrate what merging is all about, we will take an example out of the NOAA weather data. It provides an awesome minute-by-minute update of Shapefiles to satisfy your desire to download data. We will look at a week's collection of weather warnings and union these warnings together, giving us the total warning area issued in this week.

A conceptual visualization of our desired results is shown here:

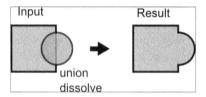

Most of the data is located around Florida, but has some polygons near Hawaii and California. To see the original data or find new data, check out these links:

▸ `http://www.nws.noaa.gov/geodata/catalog/wsom/html/pubzone.htm`

▸ `http://nws.noaa.gov/regsci/gis/week.html`

▸ `http://www.nws.noaa.gov/geodata/index.html`

If you want to see the state boundaries, you can find them at `https://www.census.gov/geo/maps-data/data/cbf/cbf_state.html`.

Here is what a sample of the data looks like around Florida before the union, which is visualized with QGIS:

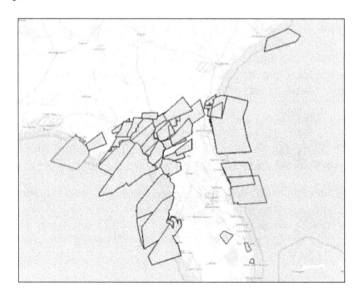

Getting ready

The usual order of business is needed to get going with this code. Fire up your virtual environment and check whether your data is all downloaded and located in your /ch06/ geodata/ folder. If all is ready, jump right in and start typing some code.

How to do it...

Our data is a little messy to say the least, so please follow our steps outlining a solution to allow us to process and run the analysis function, union:

```python
# #!/usr/bin/env python
# -*- coding: utf-8 -*-
from shapely.geometry import MultiPolygon
from shapely.ops import cascaded_union
from os.path import realpath
from utils import create_shapes
from utils import out_geoj
from utils import write_wkt


def check_geom(in_geom):
    """
    :param in_geom: input valid Shapely geometry objects
    :return: Shapely MultiPolygon cleaned
    """
    plys = []
    for g in in_geom:
        # if geometry is NOT valid
        if not g.is_valid:
            print "Oh no invalid geometry"
            # clean polygon with buffer 0 distance trick
            new_ply = g.buffer(0)
            print "now lets make it valid"
            # add new geometry to list
            plys.append(new_ply)
        else:
            # add valid geometry to list
            plys.append(g)
    # convert new polygons into a new MultiPolygon
    out_new_valid_multi = MultiPolygon(plys)
    return out_new_valid_multi


if __name__ == "__main__":
```

```
# input NOAA Shapefile
shp = realpath("../geodata/temp-all-warn-week.shp")

# output union_dissolve results as GeoJSON
out_geojson_file = realpath("../geodata/
ch06-03_union_dissolve.geojson")

out_wkt_js = realpath("ol3/data/
ch06-03_results_union.js")

# input Shapefile and convert to Shapely geometries
shply_geom = create_shapes(shp)

# Check the Shapely geometries if they are valid if not
fix them
new_valid_geom = check_geom(shply_geom)

# run our union with dissolve
dissolve_result = cascaded_union(new_valid_geom)

# output the resulting union dissolved polygons to GeoJSON
file
out_geoj(dissolve_result, out_geojson_file)

write_wkt(out_wkt_js, dissolve_result)
```

Your resulting web map will look like this:

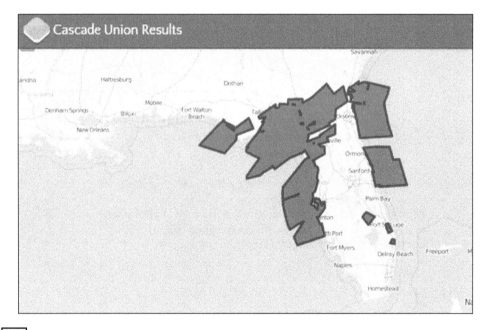

How it works...

We are starting to increasingly reuse more code that is now tucked away in our /ch06/code/ utils.py module. As you see in the imports, we use three functions for the standard input and output of data. The main application starts with defining our NOAA input Shapefile and defining the output GeoJSON file. Then, if we run the code, it will crash due to data validity issues. So, we create a new function to check our input data for invalid geometries. This new function will catch these invalid geometries and convert them to valid polygons.

Shapely has a geometry property called is_valid, which accesses the GEOS engine to check for geometry validity based on the simple features in the OGC specification.

> If you are looking for all the possible invalid data possibilities, you can find more information on the Open Geospatial Consortium website. Check out the Simple Features Standard on page *28*; you will find the examples of invalid polygons at http://portal. opengeospatial.org/files/?artifact_id=25355.

The reason for these anomalies is that when data is overlaid and processed, geometries become combined or cut at angles that are not always optimal.

At last, we have clean data to work with, making the rest of our journey very simple by running the Shapely cascaded_union() function, which will dissolve all our overlapping polygons. Our resulting MultiPolygons are pushed further into our out_geoj() function, which finally writes the new geometries to disk in our /ch06/geodata folder.

Performing an identity function (difference + intersection)

In ESRI geoprocessing terminology, there is an overlay function called identity. This is a very useful function to call when you want to keep all the original geometry boundaries of ONLY the input features combined with an intersection of input features.

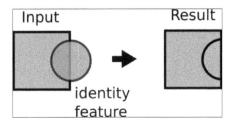

This boils down to a formula that calls for both `difference` and `intersect`. We first find the difference (`input feature - intersection`), then add the intersection to create our results as follows:

```
(input feature - intersection) + intersection = result
```

How to do it...

1. For all you curious folks who want to learn how to do this, type out the following code; it will help your muscle memory:

```python
##!/usr/bin/env python
# -*- coding: utf-8 -*-
from shapely.geometry import asShape, MultiPolygon
from utils import shp2_geojson_obj, out_geoj, write_wkt
from os.path import realpath


def create_polys(shp_data):
    """
    :param shp_data: input GeoJSON
    :return: MultiPolygon Shapely geometry
    """
    plys = []
    for feature in shp_data['features']:
        shape = asShape(feature['geometry'])
        plys.append(shape)

    new_multi = MultiPolygon(plys)
    return new_multi


def create_out(res1, res2):
    """

    :param res1: input feature
    :param res2: identity feature
    :return: MultiPolygon identity results
    """
    identity_geoms = []

    for g1 in res1:
        identity_geoms.append(g1)
    for g2 in res2:
        identity_geoms.append(g2)
```

```python
    out_identity = MultiPolygon(identity_geoms)
    return out_identity


if __name__ == "__main__":
    # out two input test Shapefiles
    shp1 = realpath("../geodata/temp1-ply.shp")
    shp2 = realpath("../geodata/temp2-ply.shp")

    # output resulting GeoJSON file
    out_geojson_file = realpath("../geodata/result_identity.
geojson")

    output_wkt_identity = realpath("ol3/data/
ch06-04_results_identity.js")


    # convert our Shapefiles to GeoJSON
    # then to python dictionaries
    shp1_data = shp2_geojson_obj(shp1)
    shp2_data = shp2_geojson_obj(shp2)

    # transform our GeoJSON data into Shapely geom objects
    shp1_polys = create_polys(shp1_data)
    shp2_polys = create_polys(shp2_data)

    # run the difference and intersection
    res_difference = shp1_polys.difference(shp2_polys)
    res_intersection = shp1_polys.intersection(shp2_polys)

    # combine the difference and intersection polygons into
results
    result_identity = create_out(res_difference,
res_intersection)

    # export identity results to a GeoJSON
    out_geoj(result_identity, out_geojson_file)

    # write out new JavaScript variable with wkt geometry
    write_wkt(output_wkt_identity, result_identity )
```

The resulting polygons can now be visualized in your browser. Now simply open the `/ch06/code/ol3/html/ch06-04_identity.html` file and you will see this map:

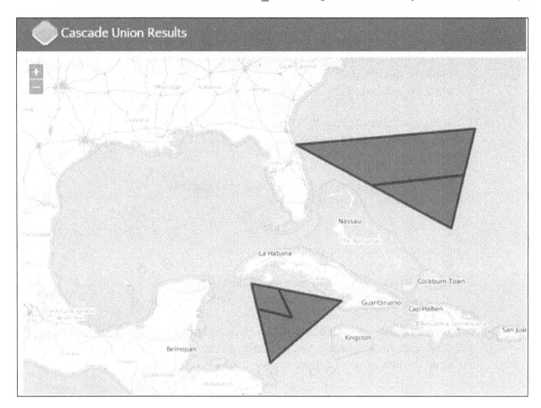

How it works...

We have hidden away a couple of gems in our `util.py` utilities file called `shp2_geojson_obj` and `out_geoj`. The first one takes in our Shapefile and returns a Python dictionary object. Our function actually creates a valid GeoJSON in the form of a Python dictionary that could very easily be converted to a JSON string using the standard `json.dumps()` Python module.

With this overhead out of the way, we can jump into creating Shapely geometries that can be used for our analysis. The `create_polys()` function does exactly this: it takes in our geometries, returning a `MultiPolygon`. This `MultiPolygon` is used to calculate our difference and intersection.

So, at last, we can do the analysis calculation starting with the Shapely difference function using our `temp1-ply.shp` as our input feature and `temp2-poly.shp` as our identity feature. The difference function only returns the geometries of the input features that do not intersect the other feature. Next up, we execute the intersection function that only returns geometries that overlap between our two inputs.

Our recipe is almost completed; we only need to combine these two new results to produce our new identity result's MultiPolygon. The `create_out()` function takes two arguments, the first being our input features and the second is our resulting intersection features. The order is very important; otherwise your results will be reversed. So make sure that you enter the correct order of input.

We run through each of the geometries and combine them into a fancy new `MultiPolygon` called `result_identity`. This is then pumped into our `out_geoj()` function, which writes out a new GeoJSON file to your `/ch06/geodata/` folder.

Our `out_geoj()` function is located in the `utils.py` file and might need a quick explanation. The input is a list of geometries and the file path of the output GeoJSON file location on disk. We simply create a new dictionary, and then loop through each geometry, exporting the Shapely geometry to a GeoJSON file using the built-in Shapely `__geo_interface__`.

 If you want to read up on the `__geo_interface__`, do so for yourself and find out what it is and why it's so cool at `https://gist.github.com/sgillies/2217756`.

For those of you looking for the two utility functions, here they are for your reading pleasure:

```
def shp2_geojson_obj(shapefile_path):
    # open shapefile
    in_ply = shapefile.Reader(shapefile_path)
    # get a list of geometry and records
    shp_records = in_ply.shapeRecords()
    # get list of fields excluding first list object
    fc_fields = in_ply.fields[1:]

    # using list comprehension to create list of field names
    field_names = [field_name[0] for field_name in fc_fields ]
    my_fc_list = []
    # run through each shape geometry and attribute
    for x in shp_records:
        field_attributes = dict(zip(field_names, x.record))
        geom_j = x.shape.__geo_interface__
        my_fc_list.append(dict(type='Feature', geometry=geom_j,
                               properties=field_attributes))

    geoj_json_obj = {'type': 'FeatureCollection',
                     'features': my_fc_list}

    return geoj_json_obj
```

```python
def out_geoj(list_geom, out_geoj_file):
    out_geojson = dict(type='FeatureCollection', features=[])

    # generate geojson file output
    for (index_num, ply) in enumerate(list_geom):
        feature = dict(type='Feature',
properties=dict(id=index_num))
        feature['geometry'] = ply.__geo_interface__
        out_geojson['features'].append(feature)

    # create geojson file on disk
    json.dump(out_geojson, open(out_geoj_file, 'w'))
```

7

Raster Analysis

In this chapter, we will cover the following topics:

- ▶ Loading a DEM USGS ACSII CDED into PostGIS
- ▶ Creating an elevation profile
- ▶ Creating a hillshade raster from your DEM with ogr
- ▶ Generating slope and aspect images from your DEM
- ▶ Merging rasters to generate a color relief map

Introduction

Raster analysis works similar to vector analysis but the spatial relation is determined by the position of the raster cell. Most of our raster data is collected through diverse remote sensing techniques. In this chapter, the goals are quite simple and focused on working with and around a **digital elevation model** (**DEM**). The DEM we are using is from Whistler, BC, Canada, home to the 2010 Winter Olympics. Our DEM is in the form of the USGS ASCII CDED (.dem) format. The DEM is our source data that is used to derive several new raster datasets. As with other chapters, we will leverage Python as our glue to run scripts to enable a processing pipeline for raster data. The visualization of our data will play out with matplotlib along with the QGIS desktop GIS.

Loading a DEM USGS ACSII CDED into PostGIS

Importing and working with a DEM in PostGIS is what this recipe is all about. We begin our journey with a text file that's full of points and is stored in the USGS ASCII CDED format (to read more about the details of this format, feel free to look at the documentation page at `http://www.gdal.org/frmt_usgsdem.html`). The ASCII format is well known and accepted by many desktop GIS applications as a direct data source. Feel free to simply open up your ASCII file with QGIS to view the files and see the resulting raster representation that it creates for you. Our task at hand is to import this DEM file into a PostGIS database, creating a new PostGIS raster dataset within PostGIS We perform this task by using a command-line tool called `raster2pgsql`, which is installed along with your PostGIS installation. The `raster2pgsql` tool is located on Windows under `C:\Program Files\ PostgreSQL\9.3\bin\` if you are running PostgreSQL 9.

Getting ready

Your data is available in the `ch07/geodata/dem_3857.dem` folder. Feel free to get the original DEM from GeoGratis Canada, the area around Whistler Mountain, British Columbia, at `http://ftp2.cits.rncan.gc.ca/pub/geobase/official/cded/50k_ dem/092/092j02.zip`.

If you have not already created your `Postgresql` database in *Chapter 1, Setting Up Your Geospatial Python Environment*, do so now and then continue with starting your virtual environment to run this script.

Also, make sure that the `raster2pgsql` command is available in your command prompt. If not, set up your environment variables on Windows or a sym link on your Linux machine.

How to do it...

Let's move on to the fun part that can be found in your `/ch07/code/ch07-01_ dem2postgis.py` file:

1. The code found in the `/ch07/code/ch07-01_dem2postgis.py` file is as follows:

    ```python
    #!/usr/bin/env python
    # -*- coding: utf-8 -*-
    import subprocess
    import psycopg2
    ```

```
db_host = "localhost"
db_user = "pluto"
db_passwd = "secret"
db_database = "py_geoan_cb"
db_port = "5432"

# connect to DB
conn = psycopg2.connect(host=db_host, user=db_user,
                        port=db_port, password=db_passwd,
                        database=db_database)

# create a cursor
cur = conn.cursor()

# input USGS ASCII DEM (and CDED)
input_dem = "../geodata/dem_3857.dem"

# create an sql file for loading into the PostGIS database
raster
# command line with options
# -c create new table
# -I option will create a spatial GiST index on the raster
column
# -C will apply raster constraints
# -M vacuum analyse the raster table

command = 'raster2pgsql -c -C -I -M ' + input_dem + '
geodata.dem_3857'

# write the output to a file

temp_sql_file = "temp_sql.sql"

# open, create new file to write sql statements into
with open(temp_sql_file, 'wb') as f:
    try:
        result = subprocess.call(command, stdout=f,
shell=True)
        if result != 0:
            raise Exception('error code %d' % result)
```

```
        except Exception as e:
            print e

    # open the file full of insert statements created by
    raster2pgsql
    with open(temp_sql_file, 'r') as r:
        # run through and execute each line inside the temp sql
    file
        for sql_insert in r:
            cur.execute(sql_insert)

    print "please open QGIS >= 2.8.x and view your loaded DEM
    data"
```

How it works...

Python, once again, is our glue that leverages the power of a command-line tool to do the dirty work. This time around, we use the Python subprocess module to call `raster2pgsql` the command-line tool. The `psycopg2` module then executes our `insert` statements.

Starting from the top and working our way down, we see the database connection settings for `psycopg2`. The input path to our DEM is set as the `input_dem` variable. Then, we pack our command-line arguments into a single string called `command`. This is then run by subprocess. The individual command-line arguments are described in the code comments and further information and options can be found directly at `http://postgis.refractions.net/docs/using_raster.xml.html#RT_Raster_Loader`.

Now that the command is ready, we need to create a temporary file to store the generated SQL `insert` and `create` statements that the `raster2pgsql` command creates. Using the `with open()` syntax, we create our temporary file and then call the command using subprocess. We use `stdout` to specify where to write out this file. The `shell=True` argument comes with a *big* warning.

The following is the `mention` warning taken from the Python documentation:

```
Warning Executing shell commands that incorporate
unsanitized input from an untrusted source makes a
program vulnerable to shell injection, a serious
security flaw which can result in arbitrary command
execution. For this reason, the use of shell=True is
strongly discouraged in cases where the command string
is constructed from external input:
```

If all goes well, no exceptions should pop up, but if they do, we catch them using the standard Python `try` statement.

The last step is to open the newly created SQL file that's full of inserts and execute each line in the file using `psycopg2`. This populates our new table that has the name of the input DEM file.

Go ahead and open up **QGIS | 2.8.x** and have a look at the raster you've just loaded into PostGIS.

> To open the raster in QGIS, I've found that you need to open the Database Manager application that comes with QGIS and connect to your Postgresql-PostGIS database and schema. Then, you will see the new raster, and you will need to right-click on it to add it to the canvas. This will finally add the raster to your QGIS project.

Creating an elevation profile

Creating an elevation profile is very helpful when trying to visualize a 3D terrain cross-section or simply to see the elevation gain of a bike tour. In this example, we will define our own LineString geometry and extract the elevation values from the DEMs that are located every 20 m along our line. The analysis will generate a new CSV file that we can open in Libre Office Calc or Microsoft Excel to visualize the new data as a line chart.

The 2D view of our line plotted on top of the elevation model as seen inside QGIS looks like this:

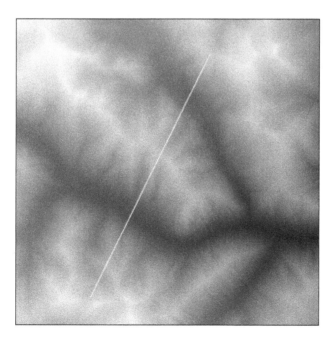

Getting ready

This recipe calls for GDAL and Shapely. Make sure that you have them installed and are running them inside your python virtual environment that you set up earlier. To visualize your final CSV file, you must also install Libre Office Calc or some other charting software. The code to execute this is located at /ch07/code/ch07-02_elev_profile.py.

How to do it...

Running the script directly from your command line will generate your CSV, so read the code comments to see all the little details of what is going on in order to generate our new file as follows:

```python
#!/usr/bin/env python
# -*- coding: utf-8 -*-
import sys, gdal, os
from gdalconst import GA_ReadOnly
from os.path import realpath
from shapely.geometry import LineString


def get_elevation(x_coord, y_coord, raster, bands, gt):
    """
    get the elevation value of each pixel under
    location x, y
    :param x_coord: x coordinate
    :param y_coord: y coordinate
    :param raster: gdal raster open object
    :param bands: number of bands in image
    :param gt: raster limits
    :return: elevation value of raster at point x,y
    """
    elevation = []
    xOrigin = gt[0]
    yOrigin = gt[3]
    pixelWidth = gt[1]
    pixelHeight = gt[5]
    px = int((x_coord - xOrigin) / pixelWidth)
    py = int((y_coord - yOrigin) / pixelHeight)
    for j in range(bands):
        band = raster.GetRasterBand(j + 1)
```

```
        data = band.ReadAsArray(px, py, 1, 1)
        elevation.append(data[0][0])
    return elevation


def write_to_csv(csv_out,result_profile_x_z):
    # check if output file exists on disk if yes delete it
    if os.path.isfile(csv_out):
        os.remove(csv_out)

    # create new CSV file containing X (distance) and Z
value pairs
    with open(csv_out, 'a') as outfile:
        # write first row column names into CSV
        outfile.write("distance,elevation" + "\n")
        # loop through each pair and write to CSV
        for x, z in result_profile_x_z:
            outfile.write(str(round(x, 2)) + ',' +
str(round(z, 2)) + '\n')


if __name__ == '__main__':
    # set directory
    in_dem = realpath("../geodata/dem_3857.dem")

    # open the image
    ds = gdal.Open(in_dem, GA_ReadOnly)

    if ds is None:
        print 'Could not open image'
        sys.exit(1)

    # get raster bands
    bands = ds.RasterCount

    # get georeference info
    transform = ds.GetGeoTransform()

    # line defining the the profile
    line = LineString([(-13659328.8483806, 6450545.73152317),
(-13651422.7820022, 6466228.25663444)])
    # length in meters of profile line
```

```
            length_m = line.length

            # lists of coords and elevations
            x = []
            y = []
            z = []
            # distance of the topographic profile
            distance = []
            for currentdistance in range(0, int(length_m), 20):
                # creation of the point on the line
                point = line.interpolate(currentdistance)
                xp, yp = point.x, point.y
                x.append(xp)
                y.append(yp)
                # extraction of the elevation value from the MNT
                z.append(get_elevation(xp, yp, ds, bands,
        transform)[0])
                distance.append(currentdistance)

            print (x)
            print (y)
            print (z)
            print (distance)

            # combine distance and elevation vales as pairs
            profile_x_z = zip(distance,z)

            csv_file = os.path.realpath('../geodata/output_profile.csv')
            # output final csv data
            write_to_csv(csv_file, profile_x_z)
```

How it works...

There are two functions that are used to create our elevation profile. The first
get_elevation() function returns a single elevation value per pixel for each band in a
raster. This means that our input raster can contain multiple bands of data. Our second
function will write our results to a CSV file.

The `get_elevation()` function creates a list of elevation values; to achieve this, we need to extract some details from our input elevation raster. The *x* and *y* origin coordinates are used in combination with the raster pixel width and height to help find pixels in our raster. This information is then processed with our input *x* and *y* coordinates where we want the elevation value to be extracted.

Next up, we loop through all the available bands in our raster and find the elevation value per band that's located at coordinates *x* and *y* from our input. The `ReadAsArray` GDAL function finds this location, and then all we need to do is get the first object of the second nested list array. This value is then appended to a new list of elevation values.

To process our data, we define the input paths of our raster with the `os.path.realpath()` Python function that returns the full path to our input. GDAL is used to open our DEM raster and return the number of bands plus the *x* origin, *y* origin, pixel width, and pixel height information from our raster. This is located in the transform variable that's passed into our `get_elevation()` function.

Working our way further, we define our input LineString. This LineString defines where the cross-section profile is going to be extracted. To process our data, we want to extract elevation values every 20 m along our input LineString. This is done in the `for` loop as we specify the range based on the LineString length and our input of 20 m. Using the Shapely `Interpolate` linear referencing function, we then create a point object every 20 m. These values are then stored in separate *x*, *y*, and *z* lists, which are then updated. The *z* list contains our list of new elevation points. Individual elevations are collected by specifying the first object in the list that's returned by our `get_elevation()` function.

To put all this together in a CSV file, we use the Python `zip` function to combine the distance values with the elevation values. This creates the final two columns of data, showing us the distance from the starting point of our LineString on the *x*-axis and the elevation value on the *y*-axis.

Visualizing the results is then easy in Libre Office Calc or Microsoft Excel. Go ahead and open up the output CSV file located in your /ch07/geodata/output_profile.csv folder and create a simple line chart:

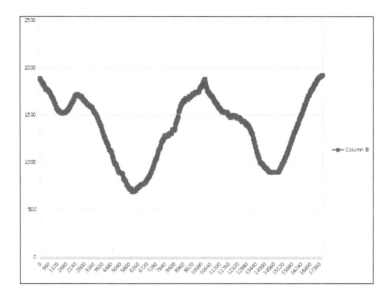

Your resulting chart should look similar to what is shown in the preceding screenshot.

To plot the graph using Libre Office Calc, see the following plotting options:

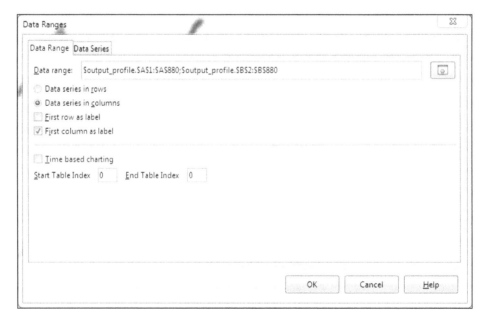

Creating a hillshade raster from your DEM with ogr

Our DEM can be the basis for many types of derived raster datasets. One of these derivatives is the so called **hillshade** raster dataset. A hillshade raster represents a 2D view of 3D elevation data, assigning gray raster shades and giving them a 3D effect by enabling you to see the highs and lows of your terrain. The hillshade is a pure visualization helper to create a nice looking map and show relief on a 2D map.

The pure Python solution to creating a hillshade is written by Roger Veciana i Rovira and you can find it at `http://geoexamples.blogspot.co.at/2014/03/shaded-relief-images-using-gdal-python.html`. There is also a nice solution by Joel Lawhead in *Chapter 7, Python and Elevation Data* of *Learning Geospatial Analysis with Python*. For those of you looking for a detailed description of the hillshade from ESRI, check this page out at `http://webhelp.esri.com/arcgisdesktop/9.3/index.cfm?TopicName=How%20Hillshade%20works`. The `gdaldem` hillshade command-line tool will be used to generate the image to disk.

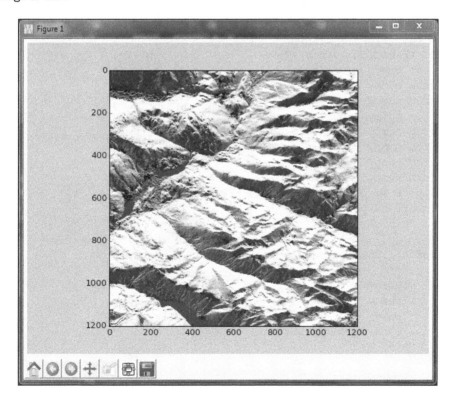

Getting ready

The prerequisites for this example require the `gdal` (`osgeo`), `numpy`, and `matplotlib` python libraries. Plus, you need to have downloaded the data folder for this book and have the `/ch07/geodata` folder available for read/write access. We are directly accessing our USGS ASCII CDED DEM `.dem` file on disk to render our hillshade, so be sure that you have this folder. The code execution will take place as usual from your `/ch07/code/` folder that runs the `ch07-03_shaded_relief.py` python file. So, for the impatient coders go ahead and give it a try at the command line as follows:

```
>> python ch07-03_shaded_relief.py
```

How to do it...

Our Python script will run through a few mathematical operations and call the gdaldem command-line tool to generate our output using the following steps:

1. The code contains some math that is not always easy to follow; the calculation of the greyscale values is determined by the elevation and its surrounding pixels, so read along:

```python
#!/usr/bin/env python
# -*- coding: utf-8 -*-
from osgeo import gdal
from numpy import gradient
from numpy import pi
from numpy import arctan
from numpy import arctan2
from numpy import sin
from numpy import cos
from numpy import sqrt
import matplotlib.pyplot as plt
import subprocess


def hillshade(array, azimuth, angle_altitude):
    """
    :param array: input USGS ASCII DEM / CDED .dem
    :param azimuth: sun position
    :param angle_altitude: sun angle
    :return: numpy array
    """

    x, y = gradient(array)
    slope = pi/2. - arctan(sqrt(x*x + y*y))
    aspect = arctan2(-x, y)
    azimuthrad = azimuth * pi / 180.
```

```
        altituderad = angle_altitude * pi / 180.


        shaded = sin(altituderad) * sin(slope)\
         + cos(altituderad) * cos(slope)\
         * cos(azimuthrad - aspect)
        return 255*(shaded + 1)/2

ds = gdal.Open('../geodata/092j02_0200_demw.dem')
arr = ds.ReadAsArray()

hs_array = hillshade(arr, 90, 45)
plt.imshow(hs_array,cmap='Greys')
plt.savefig('../geodata/hillshade_whistler.png')
plt.show()

# gdal command line tool called gdaldem
# link  http://www.gdal.org/gdaldem.html
# usage:
# gdaldem hillshade input_dem output_hillshade
# [-z ZFactor (default=1)] [-s scale* (default=1)]"
# [-az Azimuth (default=315)] [-alt Altitude (default=45)]
# [-alg ZevenbergenThorne] [-combined]
# [-compute_edges] [-b Band (default=1)] [-of format] [-co
"NAME=VALUE"]* [-q]


create_hillshade = '''gdaldem hillshade -az 315 -alt 45 ../
geodata/092j02_0200_demw.dem ../geodata/hillshade_3857.tif'''

subprocess.call(create_hillshade)
```

How it works...

The hillshade function calculates slope and aspect values for each cell as the input to calculate the shaded gray value. The `azimuth` variable defines the direction of light in degrees that hits our DEM. Inverting and playing with `azimuth` can lead to some effects, such as valleys looking like hills and hills looking like valleys. Our `shaded` variable holds the shade values as an array that we can plot with matplotlib.

Using the `gdaldem` command-line tool is definitely more robust and faster than the pure Python solution. With `gdaldem`, we create a new hillshade TIF file on disk that can open either with a local image viewer or can be drag-and-dropped into QGIS. QGIS will automatically stretch the gray values for you so that you will be able to see a nice representation of your hillshade.

Generating slope and aspect images from your DEM

Slope maps are very useful, for example, to help biologists identify habitat zones. Certain species only live in areas that are very steep—mountain goats, for instance. The slope raster can then quick identify potential habitat areas. To visualize this, we use QGIS to display our slope map, which will look similar to the following image. The area in white indicates the steeper area and the darker the color, the flatter the terrain:

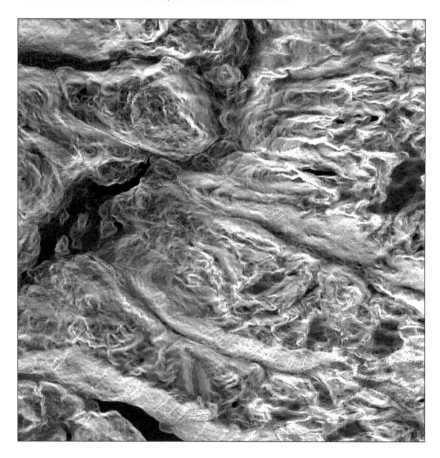

Our aspect map displays the direction that the surface faces towards—such as north, east, south, and west—and this is expressed in degrees. In the screenshot, the orange area represents warm south-facing areas. The north-facing sides are cooler and are indicated in different hues of blues from our color spectrum. To achieve the colors, the QGIS singleband pseudocolor was classified into five continuous classes as shown in the following screenshot:

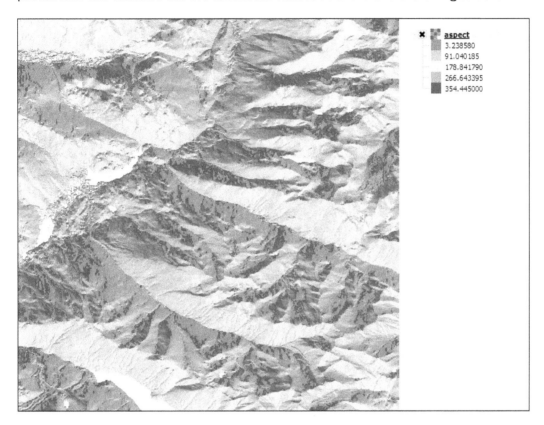

Getting ready

Ensure that your /ch07/geodata folder is downloaded and the DEM 092j02_0200_demw. dem file from Whistler, BC, Canada, is available.

How to do it...

1. We utilize the `gdaldem` command-line tool to create our slope raster. You can tweak this recipe to batch generate slope images from several DEM rasters.

```python
#!/usr/bin/env python
# -*- coding: utf-8 -*-
import subprocess


# SLOPE
# - To generate a slope map from any GDAL-supported
elevation raster :
# gdaldem slope input_dem output_slope_map"
# [-p use percent slope (default=degrees)]
[-s scale* (default=1)]
# [-alg ZevenbergenThorne]
# [-compute_edges] [-b Band (default=1)] [-of format]
[-co "NAME=VALUE"]* [-q]

create_slope = '''gdaldem slope ../geodata/092j02_0200_demw.dem
../geodata/
slope_w-degrees.tif '''

subprocess.call(create_slope)


# ASPECT
# - To generate an aspect map from any GDAL-supported
elevation raster
# Outputs a 32-bit float raster with pixel values from
0-360 indicating azimuth :
# gdaldem aspect input_dem output_aspect_map"
# [-trigonometric] [-zero_for_flat]
# [-alg ZevenbergenThorne]
# [-compute_edges] [-b Band (default=1)] [-of format]
[-co "NAME=VALUE"]* [-q]

create_aspect = '''gdaldem aspect ../geodata/092j02_0200_demw.dem
../geodata/aspect_w.tif '''

subprocess.call(create_aspect)
```

How it works...

The `gdaldem` command-line tool is our workhorse once again and all we need to do is pass along our DEM and specify an output file. Inside the code, you'll see the arguments passed include `-co compress=lzw`, which reduces the size of the image dramatically. Our `-p` option states that we want the results in a percentage slope followed by the input DEM and our output file.

As for our `gdaldem` aspect raster, the same compression applies this time and no other arguments are needed to generate the aspect raster. To visualize the aspect raster, open it inside QGIS and assign it a color as described earlier in the introduction.

Merging rasters to generate a color relief map

Generating a color relief raster is a one-liner with the `gdaldem color-relief` command line. If you want something that's a little more visually appealing, we will perform a combination of a slope, hillshade, and some color relief. Our end result is a single new raster representing a merge of layers to give a nice visual effect of elevation relief. The results are going to look similar to the following image:

Getting ready

For this exercise, you need to have the GDAL libraries installed that include the `gdaldem` command-line tool.

How to do it...

1. Let's begin by extracting some key information out of our DEM using the `gdalinfo` `\ch07\code>gdalinfo ../geodata/092j02_0200_demw.dem` command-line tool as follows:

```
Driver: USGSDEM/USGS Optional ASCII DEM (and CDED)
Files: ../geodata/092j02_0200_demw.dem
       ../geodata/092j02_0200_demw.dem.aux.xml
Size is 1201, 1201
Coordinate System is:
GEOGCS["NAD83",
    DATUM["North_American_Datum_1983",
        SPHEROID["GRS 1980",6378137,298.257222101,
            AUTHORITY["EPSG","7019"]],
        TOWGS84[0,0,0,0,0,0,0],
        AUTHORITY["EPSG","6269"]],
    PRIMEM["Greenwich",0,
        AUTHORITY["EPSG","8901"]],
    UNIT["degree",0.0174532925199433,
        AUTHORITY["EPSG","9108"]],
    AUTHORITY["EPSG","4269"]]
Origin = (-123.000104166666630,50.250104166666667)
Pixel Size = (0.000208333333333,-0.000208333333333)
Metadata:
  AREA_OR_POINT=Point
Corner Coordinates:
Upper Left  (-123.0001042,  50.2501042) (123d 0' 0.37"W,
50d15' 0.38"N)
Lower Left  (-123.0001042,  49.9998958) (123d 0' 0.37"W,
49d59'59.63"N)
Upper Right (-122.7498958,  50.2501042) (122d44'59.62"W,
50d15' 0.38"N)
Lower Right (-122.7498958,  49.9998958) (122d44'59.62"W,
49d59'59.63"N)
Center      (-122.8750000,  50.1250000) (122d52'30.00"W,
50d 7'30.00"N)
```

```
Band 1 Block=1201x1201 Type=Int16, ColorInterp=Undefined
  Min=348.000 Max=2885.000
  Minimum=348.000, Maximum=2885.000, Mean=1481.196, StdDev=564.262
  NoData Value=-32767
  Unit Type: m
  Metadata:
    STATISTICS_MAXIMUM=2885
    STATISTICS_MEAN=1481.1960280116
    STATISTICS_MINIMUM=348
    STATISTICS_STDDEV=564.26229690401
```

2. This key information is then used to create our color `ramp.txt` file. Start off by creating a new text file called `ramp.txt` and type in the following color codes:

```
-32767 255 255 255
0 46 154 88
360 251 255 128
750 96 108 31
1100 148 130 55
2900 255 255 255
```

3. The `-32767` value defines our NODATA values in the white (255 255 255) RGB color. Now, save the `ramp.txt` file in the same folder as the following code that will generate the new raster color relief:

```python
#!/usr/bin/env python
# -*- coding: utf-8 -*-
import subprocess

dem_file = '../geodata/092j02_0200_demw.dem'
hillshade_relief = '../geodata/hillshade.tif'
relief = '../geodata/relief.tif'
final_color_relief = '../geodata/final_color_relief.tif'

create_hillshade = 'gdaldem hillshade -co compress=lzw
-compute_edges ' + dem_file +  ' ' + hillshade_relief
subprocess.call(create_hillshade, shell=True)
print create_hillshade

cr = 'gdaldem color-relief -co compress=lzw ' + dem_file +
' ramp.txt ' + relief
subprocess.call(cr)
print cr
```

```
merge = 'python hsv_merge.py ' + relief + ' ' +
hillshade_relief + ' ' + final_color_relief
subprocess.call(merge)
print merge

create_slope = '''gdaldem slope -co compress=lzw
../geodata/092j02_0200_demw.dem ../geodata/slope_w-degrees.tif '''

subprocess.call(create_slope)
```

How it works...

We need to chain together some commands and variables to get the desired results to look good. To begin our journey, we'll extract some key information from our DEM to enable us to create a color ramp that defines what colors are assigned to the elevation values. This new ramp.txt file stores our color ramp values that are then used by the gdaldem color-relief command.

The code then begins by defining the input and output needed as variables throughout this script. In the preceding code we have defined the input DEM and three output .tif files.

The first call will execute the gdaldem hillshade command to generate our hillshade. This is closely followed by the gdaldem color-relief command, creating our nice color raster that's based on the ramp.txt file we've defined. The ramp.txt file contains the NODATA value and sets it as the white RGB color. The five categories are based on the DEM data itself.

The final merge takes place using the Frank Warmerdam hsv_merge.py script that combines our relief output with the generated hillshade raster, leaving us with our final raster. Our result is a nice looking combination of a color-relief map and a hillshade.

8

Network Routing Analysis

In this chapter, we will cover the following topics:

- ▶ Finding the Dijkstra shortest path with pgRouting
- ▶ Finding the Dijkstra shortest path with NetworkX in pure Python
- ▶ Generating evacuation polygons based on an indoor shortest path
- ▶ Creating centerlines from polygons
- ▶ Building an indoor routing system in 3D
- ▶ Calculating indoor route walk time

Introduction

Routing has become commonplace on navigation devices for road networks across the world. If you want to know how to drive from point A to point B, simply enter the start address and end address into your navigation software and it will calculate the shortest route for you in seconds.

Here's a scenario you may come across: Route me to Prof. Dr. Smith's office in the Geography Department for my meeting at any university anywhere. Hmm, sorry, there's no routing network available on my navigation software. This is a reminder for you to not to forget to ask for directions on campus for your meeting location.

This chapter is all about routing and, specifically, routing inside large building complexes from office *A33*, first floor in building *E01* to office *B55*, sixth floor in building *P7*.

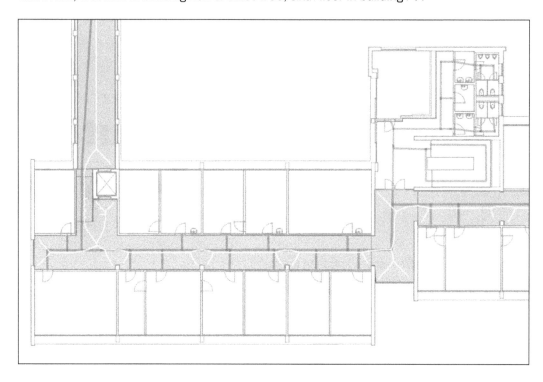

We will explore the powerful routing capabilities of **pgRouting**, an extension of PostgreSQL. With pgRouting, we can calculate the shortest path using either the Dijkstra, A*, and/or K shortest path algorithms. Alongside pgRouting, we will use a pure Python solution with the NetworkX library to generate a route from the same source data.

> BIG IMPORTANT NOTE. Pay attention to the input network dataset used and make sure that it is in the EPSG: 3857 coordinate system, a geometric Cartesian meter system. Routing calculations using world coordinates in EPSG: 4326 must be converted if used by such a system. Also, note that the GeoJSON coordinate system is interpreted by QGIS as EPSG:4326 even though the coordinates are stored in EPSG:3857!

Finding the Dijkstra shortest path with pgRouting

There are a few Python libraries out there, such as **networkX** and **scikit-image**, that can find the shortest path over a raster or NumPy array. We want to focus on routing over a vector source and returning a vector dataset; therefore, pgRouting is a natural choice for us. Custom Python *Dijkstra* or the *A Star (A*)* shortest path algorithms exist but one that performs well on large networks is hard to find. The `pgRouting` extension of PostgreSQL is used by OSM and many other projects and is well tested.

Our example will have us load a Shapefile of an indoor network from one floor for simplicity's sake. An indoor network is comprised of network lines that go along the hallways and open walkable spaces within a building, leading to a door in most cases.

Getting ready

For this recipe, we are going to need to set up our PostGIS database with the pgRouting extension. On a Windows machine, you can install pgRouting by downloading a ZIP file for Postgresql 9.3 at `http://winnie.postgis.net/download/windows/pg93/buildbot/`. Then, extract the zip file into `C:\Program Files\PostgreSQL\9.3\`.

For Ubuntu Linux users, the pgRouting website explains the details at `http://docs.pgrouting.org/2.0/en/doc/src/installation/index.html#ubuntu-debian`.

To enable this extension, you have a couple of options. First off, you can run the command-line `psql` tool to activate the extension as follows if you have your PostgreSQL running as explained in *Chapter 1*, *Setting Up Your Geospatial Python Environment*:

```
> psql py_geoan_cb -c "create extension pgrouting"
```

You can use the **pgAdmin** user tool by simply opening up the `py_geoan_cb` database, right-clicking on **Extensions**, selecting **New Extension...**, and in the **Name** field, scrolling down to find the `pgRouting` entry and selecting it.

Now we need some data to do our routing calculations. The data used is a Shapefile located in your `/ch08/geodata/shp/e01_network_lines_3857.shp` folder. Take a look at *Chapter 3*, *Moving Spatial Data from One Format to Another*, on how to import the Shapefile or use `shp2pgsql`. Here is the command-line one-liner using `ogr2ogr` to import the Shapefile:

```
>ogr2ogr -a_srs EPSG:3857 -lco "SCHEMA=geodata" -lco "COLUMN_
TYPES=type=varchar,type_id=integer" -nlt MULTILINESTRING -nln ch08_
e01_networklines -f PostgreSQL "PG:host=localhost port=5432 user=pluto
dbname=py_geoan_cb password=secret" geodata/shp/e01_network_lines_3857.
shp
```

Note that you either use the same username and password from *Chapter 1, Setting Up Your Geospatial Python Environment*, or your own defined username and password.

For Windows users, you might need to insert the full path of your Shapefile, something that could look like `c:\somepath\geodata\shp\e01_network_lines.shp`. We explicitly set the input of the EPSG:3857 Web Mercator because, sometimes, ogr2ogr guesses the wrong projection and in this way, it ensures that it is correct on upload. Another thing to note is that we also explicitly define the output table column types because `ogr2ogr` uses numeric fields for our integers and this does not go well with `pgRouting`, so we explicitly pass the comma-separated list of field names and field types.

 For a detailed description of how ogr2ogr works, visit `http://gdal.org/ogr2ogr.html`.

Our new table includes two fields, one called `type` and the other, `type_id`. The `type_id` variable will store an integer used to identify what kind of network segment we are on, such as stairs, an indoor way, or elevator. The remaining fields are necessary for `pgRouting`, which is installed as shown in the following code, and include columns called `source`, `target`, and `cost`. The `source` and `target` columns both need to be integers, while the `cost` field is of a double precision type. These types are the requirements of the pgRouting functions.

Let's go ahead and add these fields now to our `ch08_e01_networklines` table with the help of some SQL queries:

```
ALTER TABLE geodata.ch08_e01_networklines ADD COLUMN source INTEGER;

ALTER TABLE geodata.ch08_e01_networklines ADD COLUMN target INTEGER;

ALTER TABLE geodata.ch08_e01_networklines ADD COLUMN cost DOUBLE
PRECISION;

ALTER TABLE geodata.ch08_e01_networklines ADD COLUMN length DOUBLE
PRECISION;

UPDATE geodata.ch08_e01_networklines set length = ST_Length(wkb_
geometry);
```

Once the network dataset has its new columns, we need to run the create topology `pgr_createTopology()` function. This function takes the name of our network dataset, a tolerance value, geometry field name, and a primary key field name. The function will create a new table of points on the LineString intersections, that is, nodes on a network that are in the same schema:

```
SELECT public.pgr_createTopology('geodata.ch08_e01_networklines',
        0.0001, 'wkb_geometry', 'ogc_fid');
```

The `pgr_createTopology` function parameters include the name of the networklines LineStrings containing our cost and type fields. The second parameter is the distance tolerance in meters followed by the name of the geometry column and our primary key unique id called `ogc_fid`.

Now that our tables and environment are set up, this allows us to actually create the shortest path called the Dijkstra route.

To run the Python code, make sure you have the `psycopg2` and `geojson` modules installed as described in *Chapter 1, Setting Up Your Geospatial Python Environment*.

How to do it...

1. Check out this code and follow along:

```python
#!/usr/bin/env python
# -*- coding: utf-8 -*-

import psycopg2
import json
from geojson import loads, Feature, FeatureCollection

db_host = "localhost"
db_user = "pluto"
db_passwd = "secret"
db_database = "py_geoan_cb"
db_port = "5432"

# connect to DB
conn = psycopg2.connect(host=db_host, user=db_user,
port=db_port,
                          password=db_passwd,
database=db_database)

# create a cursor
cur = conn.cursor()

start_x = 1587927.204
start_y = 5879726.142
end_x = 1587947.304
end_y = 5879611.257

# find the start node id within 1 meter of the given
coordinate
# used as input in routing query start point
start_node_query = """
```

```
    SELECT id FROM geodata.ch08_e01_networklines_vertices_pgr AS p
    WHERE ST_DWithin(the_geom, ST_GeomFromText('POINT(
%s %s)',3857), 1);"""

# locate the end node id within 1 meter of the given
coordinate
end_node_query = """
    SELECT id FROM
geodata.ch08_e01_networklines_vertices_pgr AS p
    WHERE ST_DWithin(the_geom, ST_GeomFromText('POINT(
%s %s)',3857), 1);
    """

# get the start node id as an integer
cur.execute(start_node_query, (start_x, start_y))
sn = int(cur.fetchone()[0])

# get the end node id as an integer
cur.execute(end_node_query, (end_x, end_y))
en = int(cur.fetchone()[0])


# pgRouting query to return our list of segments
representing
# our shortest path Dijkstra results as GeoJSON
# query returns the shortest path between our start and end
nodes above
# using the python .format string syntax to insert a
variable in the query
routing_query = '''
    SELECT seq, id1 AS node, id2 AS edge,
ST_Length(wkb_geometry) AS cost,
        ST_AsGeoJSON(wkb_geometry) AS geoj
      FROM pgr_dijkstra(
        'SELECT ogc_fid as id, source, target,
st_length(wkb_geometry) as cost
        FROM geodata.ch08_e01_networklines',
        {start_node},{end_node}, FALSE, FALSE
      ) AS dij_route
      JOIN  geodata.ch08_e01_networklines AS input_network
      ON dij_route.id2 = input_network.ogc_fid ;
  '''.format(start_node=sn, end_node=en)


# run our shortest path query
```

```
cur.execute(routing_query)

# get entire query results to work with
route_segments = cur.fetchall()

# empty list to hold each segment for our GeoJSON output
route_result = []

# loop over each segment in the result route segments
# create the list for our new GeoJSON
for segment in route_segments:
    geojs = segment[4]
    geojs_geom = loads(geojs)
    geojs_feat = Feature(geometry=geojs_geom,
properties={'nice': 'route'})
    route_result.append(geojs_feat)

# using the geojson module to create our GeoJSON Feature
Collection
geojs_fc = FeatureCollection(route_result)

# define the output folder and GeoJSON file name
output_geojson_route =
"../geodata/ch08_shortest_path_pgrouting.geojson"



# save geojson to a file in our geodata folder
def write_geojson():
    with open(output_geojson_route, "w") as geojs_out:
        geojs_out.write(json.dumps(geojs_fc))



# run the write function to actually create the GeoJSON
file
write_geojson()

# clean up and close database curson and connection
cur.close()
conn.close()
```

2. The resulting query, if you ran it inside `pgAdmin`, for example, would return the following:

	seq integer	node integer	edge integer	cost double precision	st_asgeojson text
1	0	1	187	6.680329309644	{"type":"MultiLineString","coordinates"
2	1	189	199	9.90822481633968	{"type":"MultiLineString","coordinates"
3	2	202	260	8.86487433724218	{"type":"MultiLineString","coordinates"
4	3	255	259	2.78737609211707	{"type":"MultiLineString","coordinates"
5	4	249	252	2.50000954175229	{"type":"MultiLineString","coordinates"
6	5	247	265	4.52459771088497	{"type":"MultiLineString","coordinates"
7	6	258	285	4.48959915931802	{"type":"MultiLineString","coordinates"
8	7	268	343	2.93661653216161	{"type":"MultiLineString","coordinates"
9	8	306	499	43.3983194100033	{"type":"MultiLineString","coordinates"
10	9	440	503	2.66199104880428	{"type":"MultiLineString","coordinates"
11	10	444	506	4.45451945998841	{"type":"MultiLineString","coordinates"
12	11	447	510	3.43284090187863	{"type":"MultiLineString","coordinates"
13	12	451	512	2.71711150557509	{"type":"MultiLineString","coordinates"
14	13	453	531	1.26469115938654	{"type":"MultiLineString","coordinates"

A route needs to be visualized on a map and not as a table. Go ahead and drag and drop your newly created `/ch08/geodata/ch08_shortest_path_pgrouting.geojson` file into QGIS. If all goes well, you should see this pretty little line, excluding the red arrows and text:

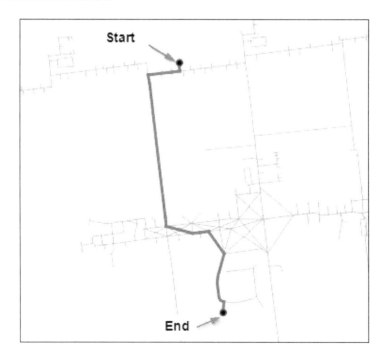

How it works...

Our code journey starts with setting up our database connection so that we can execute some queries against our uploaded data.

Now we are ready to run some routing, but wait, How do we set the start and end points that we want to route to and from? The natural way to do this is to input and the x, y coordinate pair for the start and end points. Unfortunately, the `pgr_dijkstra()` function takes only the start and end node IDs. This means that we need to get these node IDs from the new table called `ch08_e01_networklines_vertices_pgr`. To locate the nodes, we use a simple PostGIS function, `ST_Within()`, to find the nearest node within one meter from the input coordinate. Inside this query, our input geometry uses the `ST_GeomFromText()` function so that you can clearly see where things go in our SQL. Now, we'll execute our query and convert the response to an integer value as our node ID. This node ID is then ready for input in the next and final query.

The routing query will return a sequence number, node, edge, cost, and geometry for each segment along our final route. The geometry created is GeoJSON using the `ST_AsGeoJSON()` PostGIS function that feeds the creation of our final GeoJSON output route.

The pgRouting `pgr_dijkstra()` function's input arguments include an SQL query, start node ID, end node ID, directed value, and a `has_rcost` Boolean value. We set the `directed` and `has_rcost` values to `False`, while passing in the `start_node` and `end_node` IDs. This query performs a `JOIN` between the generated route IDs and input network IDs so that we have some geometry output to visualize.

Our journey then ends with processing the results and creating our output GeoJSON file. The routing query has returned a list of individual segments from start to end that aren't in the form of a single LineString, but a set of many LineStrings. This is why we need to create a list and append each route segment to a list by creating our GeoJSON `FeatureCollection` file.

Here, we use the `write_geojson()` function to output our final GeoJSON file called `ch08_shortest_path_pgrouting.geojson`.

> Note that this GeoJSON file is in the EPSG:3857 coordinate system and is interpreted by QGIS as EPSG:4326, which is incorrect. Geodata for routing, such as OSM data and custom datasets, has lots of possible mistakes, errors, and inconsistencies. Beware that the devil is hiding in the detail of the data this time and not so much in the code.

Go ahead and drag and drop your GeoJSON file into QGIS to see how your final route looks.

Finding the Dijkstra shortest path with NetworkX in pure Python

This recipe is a pure Python solution to calculate the shortest path on a network. **NetworkX** is the library we will use with many algorithms to solve the shortest path problem, including Dijkstra (http://networkx.github.io/). **NetworkX** relies on numpy and scipy to perform some graph calculations and help with performance. In this recipe, we will only use Python libraries to create our shortest path based on the same input Shapefile used in our previous recipe.

Getting ready

Start with installing *NetworkX* on your machine with the pip installer as follows:

```
>> pip install networkx
```

For the network graph algorithms, NetworkX requires numpy and scipy, so take a look at *Chapter 1*, *Setting Up Your Geospatial Python Environment*, for instructions on these. We also use Shapely to generate our geometry outputs to create GeoJSON files, so check whether you have installed Shapely. One hidden requirement is that GDAL/OGR is used in the back end of NetworkX's import Shapefile function. As mentioned earlier, in *Chapter 1,* you will find instructions on this subject.

The input data that represents our network is a Shapefile at /ch08/geodata/shp/e01_network_lines_3857.shp, containing our network dataset that is already prepared for routing, so make sure you download this chapter. Now you are ready to run the example.

How to do it...

1. You need to run this code from the command line to generate the resulting output GeoJSON files that you can open in QGIS, so follow along:

```python
#!/usr/bin/env python
# -*- coding: utf-8 -*-
import networkx as nx
import numpy as np
import json
from shapely.geometry import asLineString, asMultiPoint


def get_path(n0, n1):
    """If n0 and n1 are connected nodes in the graph,
    this function will return an array of point
    coordinates along the line linking
    these two nodes."""
```

```
        return np.array(json.loads(nx_list_subgraph[n0][n1]['Json'])[
'coordinates'])


def get_full_path(path):
    """
    Create numpy array line result
    :param path: results of nx.shortest_path function
    :return: coordinate pairs along a path
    """
    p_list = []
    curp = None
    for i in range(len(path)-1):
        p = get_path(path[i], path[i+1])
        if curp is None:
            curp = p
        if np.sum((p[0]-curp)**2) > np.sum((p[-1]-
curp)**2):
            p = p[::-1, :]
        p_list.append(p)
        curp = p[-1]
    return np.vstack(p_list)


def write_geojson(outfilename, indata):
    """
    create GeoGJSOn file
    :param outfilename: name of output file
    :param indata: GeoJSON
    :return: a new GeoJSON file
    """

    with open(outfilename, "w") as file_out:
        file_out.write(json.dumps(indata))


if __name__ == '__main__':

    # use Networkx to load a Noded shapefile
    # returns a graph where each node is a coordinate pair
    # and the edge is the line connecting the two nodes

    nx_load_shp =
nx.read_shp("../geodata/shp/e01_network_lines_3857.shp")

    # A graph is not always connected, so we take the
largest connected subgraph
```

```
        # by using the connected_component_subgraphs function.
        nx_list_subgraph = list(nx.connected_component_subgraphs(
    nx_load_shp.to_undirected()))[0]

        # get all the nodes in the network
        nx_nodes = np.array(nx_list_subgraph.nodes())

        # output the nodes to a GeoJSON file to view in QGIS
        network_nodes = asMultiPoint(nx_nodes)
        write_geojson("../geodata/ch08_final_netx_nodes.geojson",
                    network_nodes.__geo_interface__)

        # this number represents the nodes position
        # in the array to identify the node
        start_node_pos = 30
        end_node_pos = 21

        # Compute the shortest path. Dijkstra's algorithm.
        nx_short_path = nx.shortest_path(nx_list_subgraph,

    source=tuple(nx_nodes[start_node_pos]),

    target=tuple(nx_nodes[end_node_pos]),
                                        weight='distance')

        # create numpy array of coordinates representing
    result path
        nx_array_path = get_full_path(nx_short_path)

        # convert numpy array to Shapely Linestring
        out_shortest_path = asLineString(nx_array_path)

        write_geojson("../geodata/ch08_final_netx_sh_path.geojson",
                    out_shortest_path.__geo_interface__)
```

How it works...

NetworkX has a nice function called `read_shp` that inputs a Shapefile directly. However, to start doing this, we need to define the `write_geojson` function to output our results as GeoJSON files. The input Shapefile is a completely connected network dataset. Sometimes, you may find that your input is not connected and this function call to `connected_component_subgraphs` finds nodes that are connected, only using these connected nodes. The inner function sets our network to `undirected`.

This function does not create a connected network dataset; this job is left for you to perform in QGIS or some other desktop GIS software. One solution is to execute this in PostgreSQL with the tools provided with the pgRouting extension.

Now, we'll generate the nodes on our network and export them to GeoJSON. This is, of course, not necessary, but it is nice to see where the nodes are on the map to debug your data. If any problems do occur in generating routes, you can visually identify them quite quickly.

Next up, we set the array position of the start and end node to calculate our route. The NetworkX `shortest_path` algorithm requires you to define the source and target nodes.

 One thing to pay attention to is the fact that the source and target are coordinate pairs within an array of points.

As nice as this array of points are, we need a path and, hence, the `get_path` and `get_full_path` functions are discussed next. Our `get_path` function takes two input nodes, that is, two pairs of coordinates, and returns a NumPy array of edge coordinates along the line. This is followed closely by the `get_full_path` function that internally uses the `get_path` function to output the complete list of all paths and coordinates along all paths.

All the edges and corresponding coordinates are then appended to a new list that needs to be combined—hence, the NumPy `vstack` function. Inside our `for` loop, we go through each path, getting the edges and coordinates to build our list that then gets concatenated together as our final NumPy array output.

Shapely was built with NumPy compatibility and, therefore, has an `asLineString()` function that can directly input a NumPy array of coordinates. Now we have the geometry of our final LineString route and can export it to GeoJSON with our function.

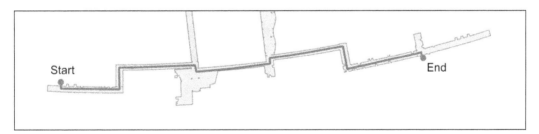

Generating evacuation polygons based on an indoor shortest path

Architects and transportation planners, for example, need to plan where and how many exits a building will require based on various standards and safety policies. After a building is built, a facility manager and security team usually do not have access to this information. Imagine that there is an event to be planned and you want to see what areas can be evacuated within a certain time, which are constrained by your list of exits in the building.

During this exercise, we want to create some polygons for a specific start point inside a major building, showing which areas can be evacuated in 10, 20, 30, and 60 second intervals. We assume that people will walk at 5 km/hr or 1.39 m/s, which is their normal walking speed. If we panic and run, our normal run speed increases to 6.7 m/s or 24.12 km/hr.

Our results are going to generate a set of polygons representing our evacuation zones based on the building hallways. We need to define the start position of where the evacuation begins. This starting point of our calculation is equal to the starting point in our route that was discussed in the previous recipe, *Finding the Dijkstra shortest path with NetworkX in pure Python*.

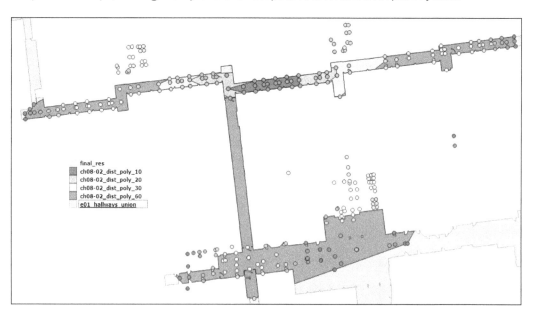

This image shows the resulting polygons and points that are generated using our script. The results are styled and visualized using QGIS.

Getting ready

This example uses the network data loaded by our previous recipe, so make sure that you have loaded this data into your local PostgreSQL database. After you have loaded the data, you will have two tables, geodata.ch08_e01_networklines_vertices_pgr and geodata.ch08_e01_networklines. In combination with these tables, you need a single new Shapefile for our input polygons located at /ch08/geodata/shp/e01_hallways_union_3857.shp, representing the building hallways that are used to clip our resulting distance polygons.

How to do it...

1. There are lots of comments in the code for clarity purposes, so read along:

```python
#!/usr/bin/env python
# -*- coding: utf-8 -*-

import psycopg2
import shapefile
import json
import shapely.geometry as geometry
from geojson import loads, Feature, FeatureCollection
from shapely.geometry import asShape

# database connection
db_host = "localhost"
db_user = "pluto"
db_passwd = "secret"
db_database = "py_geoan_cb"
db_port = "5432"

# connect to DB
conn = psycopg2.connect(host=db_host, user=db_user,
port=db_port,
                        password=db_passwd,
database=db_database)
cur = conn.cursor()

def write_geojson(outfilename, indata):
    with open(outfilename, "w") as geojs_out:
        geojs_out.write(json.dumps(indata))
```

```
# center point for creating our distance polygons
x_start_coord = 1587926.769
y_start_coord = 5879726.492


# query including two variables for the x, y POINT
coordinate
start_node_query = """
    SELECT id
    FROM geodata.ch08_e01_networklines_vertices_pgr AS p
    WHERE ST_DWithin(the_geom,
      ST_GeomFromText('POINT({0} {1})',3857),1);
      """.format(x_start_coord, y_start_coord)

# get the start node id as an integer
# pass the variables
cur.execute(start_node_query)
start_node_id = int(cur.fetchone()[0])

combined_result = []

hallways = shapefile.Reader("../geodata/shp/
e01_hallways_union_3857.shp")
e01_hallway_features = hallways.shape()
e01_hallway_shply = asShape(e01_hallway_features)

# time in seconds
evac_times = [10, 20, 30, 60]


def generate_evac_polys(start_node_id, evac_times ):
    """

    :param start_node_id: network node id to start from
    :param evac_times: list of times in seconds
    :return: none, generates GeoJSON files
    """

    for evac_time in evac_times:

        distance_poly_query = """
            SELECT seq, id1 AS node, cost,
ST_AsGeoJSON(the_geom)
                FROM pgr_drivingDistance(
```

```
                            'SELECT ogc_fid AS id, source,
target,
                                ST_Length(wkb_geometry)/
5000*60*60 AS cost
                            FROM
geodata.ch08_e01_networklines',
                            {0}, {1}, false, false
                    ) as ev_dist
                    JOIN
geodata.ch08_e01_networklines_vertices_pgr
                    AS networklines
                    ON ev_dist.id1 = networklines.id;
                """.format(start_node_id, evac_time)

        cur.execute(distance_poly_query)
        # get entire query results to work with
        distance_nodes = cur.fetchall()

        # empty list to hold each segment for our GeoJSON
output
        route_results = []

        # loop over each segment in the result route
segments
        # create the list of our new GeoJSON
        for dist_node in distance_nodes:
            sequence = dist_node[0]      # sequence number
            node = dist_node[1]          # node id
            cost = dist_node[2]          # cost value
            geojs = dist_node[3]         # geometry
            geojs_geom = loads(geojs) # create geojson geom
            geojs_feat = Feature(geometry=geojs_geom,
                    properties={'sequence_num': sequence,
                    'node':node, 'evac_time_sec':cost,
                    'evac_code': evac_time})
            # add each point to total including all points
            combined_result.append(geojs_feat)
            # add each point for individual evacuation time
            route_results.append(geojs_geom)

        # geojson module creates GeoJSON Feature Collection
        geojs_fc = FeatureCollection(route_results)

        # create list of points for each evac time
```

```
        evac_time_pts = [asShape(route_segment) for
route_segment in route_results]

        # create MultiPoint from our list of points for
evac time
        point_collection =
geometry.MultiPoint(list(evac_time_pts))

        # create our convex hull polyon around evac time
points
        convex_hull_polygon = point_collection.convex_hull

        # intersect convex hull with hallways polygon (
ch = convex hull)
        cvex_hull_intersect =
e01_hallway_shply.intersection(convex_hull_polygon)

        # export convex hull intersection to geojson
        cvex_hull = cvex_hull_intersect.__geo_interface__

        # for each evac time we create a unique GeoJSON
polygon
        output_ply = "../geodata/ch08-03_dist_poly_" +
str(evac_time) + ".geojson"

        write_geojson(output_ply, cvex_hull)

        output_geojson_route = "../geodata/
ch08-03_dist_pts_" + str(evac_time) + ".geojson"

        # save GeoJSON to a file in our geodata folder
        write_geojson(output_geojson_route, geojs_fc )

# create or set of evac GeoJSON polygons based
# on location and list of times in seconds
generate_evac_polys(start_node_id, evac_times)

# final result GeoJSON
final_res = FeatureCollection(combined_result)

# write to disk
write_geojson("../geodata/ch08-03_final_dist_poly.geojson",
final_res)

# clean up and close database cursor and connection
cur.close()
conn.close()
```

How it works...

The code starts with database boiler plate code plus a function to export the GeoJSON result files. To create an evacuation polygon, we require one input, which is the starting point for the distance calculation polygon on our network. As seen in the previous section, we need to find the node on the network closest to our starting coordinate. Therefore, we run a SQL `select` to find this node that's within one meter of our coordinate.

Next up, we define the `combined_result` variable that will hold all the points reachable for all specified evacuation times in our list. Hence, it stores the results of each evacuation time in one single output.

The hallways Shapefile is then prepared as Shapely geometry because we will need it to clip our output polygons to be inside the hallways. We are only interested in seeing which areas can be evacuated within the specified time scales of 10, 20, 30, and 60 seconds. If the area is outside the hallways, you are located outside the building and, well, better said, you are safe.

Now, we will loop through each of our time intervals to create individual evacuation polygons for each time defined in our list. The pgRouting extension includes a function called `pgr_drivingDistance()`, which returns a list of nodes that are reachable within a specified cost. Parameters for this function include the *SQL query* that returns `id`, `source`, `target`, and `cost` columns. Our final four parameters include the start node ID that's represented by the `%s` variable and equals `start_node_id`. Then, the evacuation time in seconds stored within the `evac_time` variable followed by two false values. These last two false values are for the directed route or reverse cost calculation, which we are not using.

 In our case, the cost is calculated as a time value in seconds based on distance. We assume that you are walking at 5 km/hr. The cost is then calculated as the segment length in meters divided by 5000 m x 60 min x 60 sec to derive a cost value. Then, we pass in the start node ID along with our specified evacuation time in seconds. If you want to calculate in minutes, simply remove one of the x 60 in the equation.

The geometry of each node is then derived through a SQL JOIN between the vertices table and the result list of nodes with node IDs. Now that we have our set of geometry of points for each node reachable within our evacuation time, it's time to parse this result. Parsing is required to create our GeoJSON output, and it also feeds the points into our combined output, the `combined_result` variable, and the individual evacuation time polygons that are created with a convex hull algorithm from Shapely.

A better or more realistic polygon could be created using alpha shapes. Alpha shapes form a polygon from a set of points, hugging each point to retain a more realistic polygon that follow the shape of the points. The convex hull simply ensures that all the points are inside the resulting polygon. For a good read on alpha shapes, check out this post by Sean Gillies at `http://sgillies.net/blog/1155/the-fading-shape-of-alpha/` and this post at `http://blog.thehumangeo.com/2014/05/12/drawing-boundaries-in-python/`.

What is included in the code is the alpha shapes module called `//ch08/code/alpha_shape.py` that you can try out with the input data points created, if you've followed along so far, to create a more accurate polygon.

Our `route_results` variable stores the GeoJSON geometry used to create individual convex hull polygons. This variable is then used to populate the list of points for each evacuation set of points. It also provides the source of our GeoJSON export, creating `FeatureCollection`.

The final calculations include using Shapely to create the convex hull polygon, immediately followed by intersecting this new convex hull polygon with our input Shapefile that represents the building hallways. We are only interested in showing areas to evacuate, which boils down to only areas inside the building, hence the intersection.

The remaining code exports our results to the GeoJSON files in your `/ch08/geodata` folder. Go ahead and open this folder and drag and drop the GeoJSON files into QGIS to visualize your new results. You will want to grab the following files:

▶ `ch08-03_dist_poly_10.geojson`
▶ `ch08-03_dist_poly_20.geojson`
▶ `ch08-03_dist_poly_30.geojson`
▶ `ch08-03_dist_poly_60.geojson`
▶ `ch08-03_final_dis_poly.geojson`

Creating centerlines from polygons

For any routing algorithm to work, we need a set of network LineStrings to perform our shortest path query on. Here, you, of course, have some options, ones that you can download to the OSM data to clean up the roads. Secondly, you could digitize your own set of network lines or, thirdly, you can try to autogenerate these lines.

The generation of this network LineString is of utmost importance and determines the quality and types of routes that we can generate. In an indoor environment, we have no roads and street names; instead, we have hallways, rooms, lounges, elevators, ramps, and stairs. These features are our roads, bridges, and highway metaphors where we want to create routes for people to walk.

How we can create basic network LineStrings from polygons that represent hallways is what we are going to show you in this recipe.

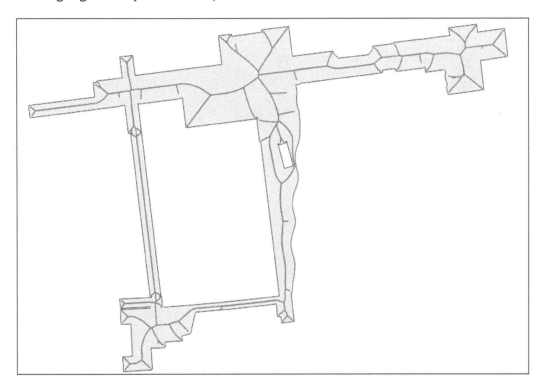

Getting ready

This exercise requires us to have a plan of some sort in digital form with polygons representing hallways and other open spaces where people could walk. Our hallway polygon is courtesy of the Alpen-Adria-Universität Klagenfurt in Austria. The polygons were simplified to keep the rendering time low. The more complex your input geometry, the longer it will take to process.

We are using the `scipy`, `shapely`, and `numpy` libraries, so read *Chapter 1, Setting Up Your Geospatial Python Environment*, if you have not done so already. Inside the `/ch08/code/` folder, you'll find the `centerline.py` module containing the `Centerline` class. This contains the actual code that generates centerlines and is imported by the `ch08/code/ch08-04_centerline.py` module.

How to do it...

Let's dive into some code:

 If you decide to run the following code straightaway, beware that the creation of centerlines is a slow process and is not optimized for performance. This code could run for 5 min on a slow machine, so be patient and keep an eye on the console until it displays **FINISHED**.

1. The first task is to create a function to create our centerlines. This is the modified version of the Filip Todic orginal `centerlines.py` class:

```python
#!/usr/bin/env python
# -*- coding: utf-8 -*-
from shapely.geometry import LineString
from shapely.geometry import MultiLineString
from scipy.spatial import Voronoi
import numpy as np


class Centerline(object):
    def __init__(self, inputGEOM, dist=0.5):
        self.inputGEOM = inputGEOM
        self.dist = abs(dist)

    def create_centerline(self):
        """
        Calculates the centerline of a polygon.

        Densifies the border of a polygon which is then
represented
        by a Numpy array of points necessary for creating
the
        Voronoi diagram. Once the diagram is created, the
ridges
        located within the polygon are joined and returned.

        Returns:
            a MultiLinestring located within the polygon.
        """

        minx =
int(min(self.inputGEOM.envelope.exterior.xy[0]))
```

```
        miny =
int(min(self.inputGEOM.envelope.exterior.xy[1]))

        border =
np.array(self.densify_border(self.inputGEOM, minx, miny))

        vor = Voronoi(border)
        vertex = vor.vertices

        lst_lines = []
        for j, ridge in enumerate(vor.ridge_vertices):
            if -1 not in ridge:
                line = LineString([
                    (vertex[ridge[0]][0] + minx,
vertex[ridge[0]][1] + miny),
                    (vertex[ridge[1]][0] + minx,
vertex[ridge[1]][1] + miny)])

                if line.within(self.inputGEOM) and
len(line.coords[0]) > 1:
                    lst_lines.append(line)

        return MultiLineString(lst_lines)

    def densify_border(self, polygon, minx, miny):
        """
        Densifies the border of a polygon by a given factor
        (by default: 0.5).

        The function tests the complexity of the polygons
        geometry, i.e. does the polygon have holes or not.
        If the polygon doesn't have any holes, its exterior
        is extracted and densified by a given factor.
        If the polygon has holes, the boundary of each hole
        as well as its exterior is extracted and densified
        by a given factor.

        Returns:
            a list of points where each point is
            represented
            by a list of its
            reduced coordinates.

        Example:
            [[X1, Y1], [X2, Y2], ..., [Xn, Yn]
```

```
        """

        if len(polygon.interiors) == 0:
            exterior_line = LineString(polygon.exterior)
            points =
self.fixed_interpolation(exterior_line, minx, miny)

        else:
            exterior_line = LineString(polygon.exterior)
            points =
self.fixed_interpolation(exterior_line, minx, miny)

            for j in range(len(polygon.interiors)):
                interior_line =
LineString(polygon.interiors[j])
                points +=
self.fixed_interpolation(interior_line, minx, miny)

        return points

    def fixed_interpolation(self, line, minx, miny):
        """
        A helping function which is used in densifying
        the border of a polygon.

        It places points on the border at the specified
        distance. By default the distance is 0.5 (meters)
        which means that the first point will be placed
        0.5 m from the starting point, the second point
        will be placed at the distance of 1.0 m from the
        first point, etc. Naturally, the loop breaks when
        the summarized distance exceeds
        the length of the line.

        Returns:
            a list of points where each point is
            represented by
            a list of its reduced coordinates.

        Example:
            [[X1, Y1], [X2, Y2], ..., [Xn, Yn]
        """

        count = self.dist
        newline = []
```

```
        startpoint = [line.xy[0][0] - minx,
line.xy[1][0] - miny]
        endpoint = [line.xy[0][-1] - minx,
line.xy[1][-1] - miny]
        newline.append(startpoint)

        while count < line.length:
            point = line.interpolate(count)
            newline.append([point.x - minx,
point.y - miny])
            count += self.dist

        newline.append(endpoint)

        return newline
```

2. Now that we have a function that creates centerlines, we need some code to import a Shapefile polygon, run the centerlines script, and export our results to GeoJSON so we that can see it in QGIS:

```python
#!/usr/bin/env python
# -*- coding: utf-8 -*-
import json
import shapefile
from shapely.geometry import asShape, mapping
from centerline import Centerline


def write_geojson(outfilename, indata):
    with open(outfilename, "w") as file_out:
        file_out.write(json.dumps(indata))


def create_shapes(shapefile_path):
    '''
    Create our Polygon
    :param shapefile_path: full path to shapefile
    :return: list of Shapely geometries
    '''
    in_ply = shapefile.Reader(shapefile_path)
    ply_shp = in_ply.shapes()

    out_multi_ply = [asShape(feature) for feature in
ply_shp]
```

```python
        print "converting to MultiPolygon: "

        return out_multi_ply


def generate_centerlines(polygon_shps):
    '''
    Create centerlines
    :param polygon_shps: input polygons
    :return: dictionary of linestrings
    '''
    dct_centerlines = {}

    for i, geom in enumerate(polygon_shps):
        print " now running Centerline creation"
        center_obj = Centerline(geom, 0.5)
        center_line_shply_line = center_obj.create_centerline()
        dct_centerlines[i] = center_line_shply_line

    return dct_centerlines


def export_center(geojs_file, centerlines):
    '''
    Write output to GeoJSON file
    :param centerlines: input dictionary of linestrings
    :return: write to GeoJSON file
    '''
    with open(geojs_file, 'w') as out:

        for i, key in enumerate(centerlines):
            geom = centerlines[key]
            newline = {'id': key, 'geometry':
mapping(geom), 'properties': {'id': key}}

            out.write(json.dumps(newline))


if __name__ == '__main__':

    input_hallways = \
"../geodata/shp/e01_hallways_small_3857.shp"
    # run our function to create Shapely geometries
```

```
shply_ply_halls = create_shapes(input_hallways)

# create our centerlines
res_centerlines = generate_centerlines(shply_ply_halls)
print "now creating centerlines geojson"

# define output file name and location
outgeojs_file =
'../geodata/04_centerline_results_final.geojson'

# write the output GeoJSON file to disk
export_center(outgeojs_file, res_centerlines)
```

How it works...

Starting with `centerlines.py` that contains the `Centerline` class, there is a lot going on inside the class. We use the **Voronoi** polygons and extract **ridges** as centerlines. To create these Voronoi polygons, we need to convert our polygon into LineStrings representing inner and outer polygon edges. These edges then need to be converted to points to feed the Voronoi algorithm. The points are generated based on a *densify* algorithm that creates points every 0.5 m along the edge of a polygon and all the way around it. This helps the `Voronoi` function create a more accurate representation of the polygon, and hence provides a better centerline. On the negative side, the higher this distance is set, the more computing power needed.

The `ch08-04_centerline.py` code then imports this new Centerline class and actually runs it using our hallways polygon. The input polygons are read from a Shapefile using `pyshp`. Our generated shapes are then pumped into the `generate_centerlines` function to output a dictionary of LineStrings representing our centerlines.

That output dictionary is then exported to GeoJSON as we loop over the centerlines and use the standard `json.dumps` function to export it to our file.

Building an indoor routing system in 3D

How to route through one or multiple buildings or floors is what this recipe is all about. This is, of course, the most complex situation involving complex data collection, preparation, and implementation processes. We cannot go into all the complex data details of collection and transformation from ACAD to PostGIS, for example; instead, the finished data is provided.

To create an indoor routing application, you need an already digitized routing network set of lines representing the areas where people can walk. Our data represents the first and second floor of a university building. The resulting indoor route, shown in the following screenshot, starts from the second floor and travels down the stairs to the first floor, all the way through the building, heading up the stairs again to the second floor, and finally reaching our destination.

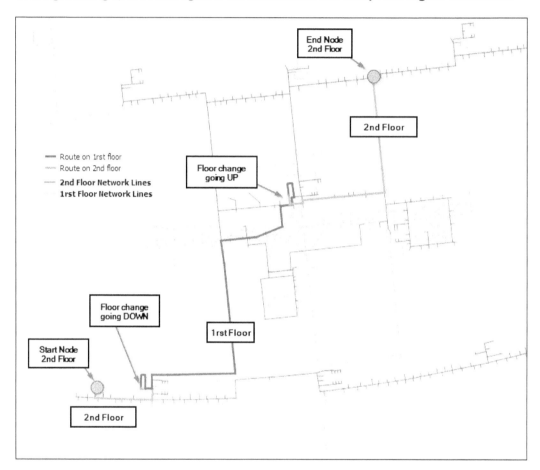

Getting ready

For this recipe, we will need to complete quite a few tasks to prepare for the indoor 3D routing. Here's a quick list of requirements:

- A Shapefile for the first floor (`/ch08/geodata/shp/e01_network_lines_3857.shp`).

- A Shapefile for the second floor (`/ch08/geodata/shp/e02_network_lines_3857.shp`).

▸ PostgreSQL DB 9.1 + PostGIS 2.1 and pgRouting 2.0. These were all installed in the *Finding the Dijkstra shortest path with pgRouting* recipe at the beginning of this chapter.

▸ Python modules, `psycopg2` and `geojson`.

Here is the list of tasks that we need to carry out:

1. Import the Shapefile of the first floor networklines (skip this if you've completed the earlier recipe that imported this Shapefile) as follows:

```
ogr2ogr -a_srs EPSG:3857 -lco "SCHEMA=geodata" -lco "COLUMN_
TYPES=type=varchar,type_id=integer" -nlt MULTILINESTRING -nln
ch08_e01_networklines -f PostgreSQL "PG:host=localhost port=5432
user=postgres dbname=py_geoan_cb password=air" geodata/shp/e01_
network_lines_3857.shp
```

2. Import the Shapefile of the second floor networklines as follows:

```
ogr2ogr -a_srs EPSG:3857 -lco "SCHEMA=geodata" -lco "COLUMN_
TYPES=type=varchar,type_id=integer" -nlt MULTILINESTRING -nln
ch08_e02_networklines -f PostgreSQL "PG:host=localhost port=5432
user=postgres dbname=py_geoan_cb password=air" geodata/shp/e02_
network_lines_3857.shp
```

3. Assign routing columns to the first floor networklines (skip this step if you've completed it in the previous recipe):

```
ALTER TABLE geodata.ch08_e01_networklines ADD COLUMN source
INTEGER;
```

```
ALTER TABLE geodata.ch08_e01_networklines ADD COLUMN target
INTEGER;
```

```
ALTER TABLE geodata.ch08_e01_networklines ADD COLUMN cost DOUBLE
PRECISION;
```

```
ALTER TABLE geodata.ch08_e01_networklines ADD COLUMN length DOUBLE
PRECISION;
```

```
UPDATE geodata.ch08_e01_networklines set length = ST_Length(wkb_
geometry);
```

4. Assign routing columns to the second floor networklines as follows:

```
ALTER TABLE geodata.ch08_e02_networklines ADD COLUMN source
INTEGER;
```

```
ALTER TABLE geodata.ch08_e02_networklines ADD COLUMN target
INTEGER;
```

```
ALTER TABLE geodata.ch08_e02_networklines ADD COLUMN cost DOUBLE
PRECISION;
```

```
ALTER TABLE geodata.ch08_e02_networklines ADD COLUMN length DOUBLE
PRECISION;

UPDATE geodata.ch08_e02_networklines set length = ST_Length(wkb_
geometry);
```

5. Create pgRouting 3D functions that allow you to route over your 3D networklines. These two PostgreSQL functions are critically important as they reflect the original pgRouting 2D functions that have now been converted to allow 3D routing. The order of installation is also very important, so make sure you install `pgr_pointtoid3d.sql` first! Both SQL files are located in your /ch08/code/ folder:

```
psql -U username -d py_geoan_cb -a -f pgr_pointtoid3d.sql
```

6. Next, install `pgr_createTopology3d.sql`. This is a modified version of the original that now uses our new `pgr_pointtoid3d` functions as follows:

```
psql -U username -d py_geoan_cb -a -f pgr_createTopology3d.sql
```

7. Now we need to merge our two floor network lines into a single 3D LineString table that we will perform our 3D routing on. This set of SQL commands is stored for you at:

```
psql -U username -d py_geoan_cb -a -f indrz_create_3d_
networklines.sql
```

The exact creation of the 3D routing table is very important to understand as it allows 3D routing queries. Our code is, therefore, listed as follows with SQL comments describing what we are doing at each step:

```
-- if not, go ahead and update
-- make sure tables dont exist


drop table if exists geodata.ch08_e01_networklines_routing;
drop table if exists geodata.ch08_e02_networklines_routing;


-- convert to 3d coordinates with EPSG:3857
SELECT ogc_fid, ST_Force_3d(ST_Transform(ST_Force_2D(st_geometryN(wkb_
geometry, 1)),3857)) AS wkb_geometry,
  type_id, cost, length, 0 AS source, 0 AS target
  INTO geodata.ch08_e01_networklines_routing
  FROM geodata.ch08_e01_networklines;


SELECT ogc_fid, ST_Force_3d(ST_Transform(ST_Force_2D(st_geometryN(wkb_
geometry, 1)),3857)) AS wkb_geometry,
  type_id, cost, length, 0 AS source, 0 AS target
```

```
  INTO geodata.ch08_e02_networklines_routing

  FROM geodata.ch08_e02_networklines;
```

```
-- fill the 3rd coordinate according to their floor number
```

```
UPDATE geodata.ch08_e01_networklines_routing SET wkb_geometry=ST_
Translate(ST_Force_3Dz(wkb_geometry),0,0,1);
```

```
UPDATE geodata.ch08_e02_networklines_routing SET wkb_geometry=ST_
Translate(ST_Force_3Dz(wkb_geometry),0,0,2);
```

```
UPDATE geodata.ch08_e01_networklines_routing SET length =ST_Length(wkb_
geometry);
```

```
UPDATE geodata.ch08_e02_networklines_routing SET length =ST_Length(wkb_
geometry);
```

```
-- no cost should be 0 or NULL/empty
```

```
UPDATE geodata.ch08_e01_networklines_routing SET cost=1 WHERE cost=0 or
cost IS NULL;
```

```
UPDATE geodata.ch08_e02_networklines_routing SET cost=1 WHERE cost=0 or
cost IS NULL;
```

```
-- update unique ids ogc_fid accordingly
```

```
UPDATE geodata.ch08_e01_networklines_routing SET ogc_fid=ogc_fid+100000;
```

```
UPDATE geodata.ch08_e02_networklines_routing SET ogc_fid=ogc_fid+200000;
```

```
-- merge all networkline floors into a single table for routing
```

```
DROP TABLE IF EXISTS geodata.networklines_3857;
```

```
SELECT * INTO geodata.networklines_3857 FROM
```

```
(
```

```
(SELECT ogc_fid, wkb_geometry, length, type_id, length*o1.cost as total_
cost,
```

```
   1 as layer FROM geodata.ch08_e01_networklines_routing o1) UNION
```

```
(SELECT ogc_fid, wkb_geometry, length, type_id, length*o2.cost as total_
cost,
```

```
    2 as layer FROM geodata.ch08_e02_networklines_routing o2))
as foo ORDER BY ogc_fid;


CREATE INDEX wkb_geometry_gist_index
   ON geodata.networklines_3857 USING gist (wkb_geometry);


CREATE INDEX ogc_fid_idx
   ON geodata.networklines_3857 USING btree (ogc_fid ASC NULLS LAST);


CREATE INDEX network_layer_idx
  ON geodata.networklines_3857
  USING hash
  (layer);


-- create populate geometry view with info
SELECT Populate_Geometry_Columns('geodata.networklines_3857'::regclass);


-- update stairs, ramps and elevators to match with the next layer
UPDATE geodata.networklines_3857 SET wkb_geometry=ST_AddPoint(wkb_
geometry,
  ST_EndPoint(ST_Translate(wkb_geometry,0,0,1)))
  WHERE type_id=3 OR type_id=5 OR type_id=7;
-- remove the second last point
UPDATE geodata.networklines_3857 SET wkb_geometry=ST_RemovePoint(wkb_
geometry,ST_NPoints(wkb_geometry) - 2)
  WHERE type_id=3 OR type_id=5 OR type_id=7;



-- add columns source and target
ALTER TABLE geodata.networklines_3857 add column source integer;
ALTER TABLE geodata.networklines_3857 add column target integer;
ALTER TABLE geodata.networklines_3857 OWNER TO postgres;


-- we dont need the temporary tables any more, delete them
DROP TABLE IF EXISTS geodata.ch08_e01_networklines_routing;
DROP TABLE IF EXISTS geodata.ch08_e02_networklines_routing;
```

```
-- remove route nodes vertices table if exists
DROP TABLE IF EXISTS geodata.networklines_3857_vertices_pgr;
-- building routing network vertices (fills source and target columns in
those new tables)
SELECT public.pgr_createTopology3d('geodata.networklines_3857', 0.0001,
'wkb_geometry', 'ogc_fid');
```

Wow, that was a lot of stuff to get through, and now we are actually ready to run and create some 3D routes. Hurray!

How to do it...

1. Let's dive into some code full of comments for your reading pleasure:

```python
#!/usr/bin/env python
# -*- coding: utf-8 -*-

import psycopg2
import json
from geojson import loads, Feature, FeatureCollection

db_host = "localhost"
db_user = "pluto"
db_passwd = "secret"
db_database = "py_geoan_cb"
db_port = "5432"

# connect to DB
conn = psycopg2.connect(host=db_host, user=db_user,
port=db_port,
                        password=db_passwd,
database=db_database)

# create a cursor
cur = conn.cursor()

# define our start and end coordinates in EPSG:3857
# set start and end floor level as integer 0,1,2 for
example
x_start_coord = 1587848.414
y_start_coord = 5879564.080
start_floor = 2

x_end_coord = 1588005.547
y_end_coord = 5879736.039
```

```
    end_floor = 2


    # find the start node id within 1 meter of the given
    coordinate
    # select from correct floor level using 3D Z value
    # our Z Value is the same as the floor number as an integer
    # used as input in routing query start point
    start_node_query = """
        SELECT id FROM geodata.networklines_3857_vertices_pgr
    AS p
        WHERE ST_DWithin(the_geom, ST_GeomFromText('POINT(
    %s %s)',3857), 1)
        AND ST_Z(the_geom) = %s;"""

    # locate the end node id within 1 meter of the given
    coordinate
    end_node_query = """
        SELECT id FROM geodata.networklines_3857_vertices_pgr
    AS p
        WHERE ST_DWithin(the_geom, ST_GeomFromText('POINT(
    %s %s)',3857), 1)
        AND ST_Z(the_geom) = %s;"""

    # run our query and pass in the 3 variables to the query
    # make sure the order of variables is the same as the
    # order in your query
    cur.execute(start_node_query, (x_start_coord,
    y_start_coord, start_floor))
    start_node_id = int(cur.fetchone()[0])

    # get the end node id as an integer
    cur.execute(end_node_query, (x_end_coord, y_end_coord,
    end_floor))
    end_node_id = int(cur.fetchone()[0])


    # pgRouting query to return our list of segments
    representing
    # our shortest path Dijkstra results as GeoJSON
    # query returns the shortest path between our start and end
    nodes above
    # in 3D traversing floor levels and passing in the layer
    value = floor

    routing_query = '''
        SELECT seq, id1 AS node, id2 AS edge,
    ST_Length(wkb_geometry) AS cost, layer,
```

```
            ST_AsGeoJSON(wkb_geometry) AS geoj
        FROM pgr_dijkstra(
            'SELECT ogc_fid as id, source, target,
st_length(wkb_geometry) AS cost, layer
            FROM geodata.networklines_3857',
            %s, %s, FALSE, FALSE
        ) AS dij_route
        JOIN  geodata.networklines_3857 AS input_network
        ON dij_route.id2 = input_network.ogc_fid ;
    '''


# run our shortest path query
cur.execute(routing_query, (start_node_id, end_node_id))

# get entire query results to work with
route_segments = cur.fetchall()

# empty list to hold each segment for our GeoJSON output
route_result = []

# loop over each segment in the result route segments
# create the list of our new GeoJSON
for segment in route_segments:
    print segment
    seg_cost = segment[3]      # cost value
    layer_level = segment[4]   # floor number
    geojs = segment[5]         # geojson coordinates
    geojs_geom = loads(geojs) # load string to geom
    geojs_feat = Feature(geometry=geojs_geom, properties={'floor':
layer_level, 'cost': seg_cost})
    route_result.append(geojs_feat)

# using the geojson module to create our GeoJSON Feature
Collection
geojs_fc = FeatureCollection(route_result)

# define the output folder and GeoJSON file name
output_geojson_route =
"../geodata/ch08_indoor_3d_route.geojson"


# save geojson to a file in our geodata folder
def write_geojson():
    with open(output_geojson_route, "w") as geojs_out:
        geojs_out.write(json.dumps(geojs_fc))
```

```
# run the write function to actually create the GeoJSON
file
write_geojson()

# clean up and close database curson and connection
cur.close()
conn.close()
```

How it works...

Using the `psycopg2` module, we can connect to our fancy new tables in the database and run some queries. The first query set finds the start and end nodes based on the x, y, and Z elevation values. The Z value is VERY important; otherwise, the wrong node will be selected. The Z value corresponds one to one with a layer/floor value. The 3D elevation data assigned to our `networklines_3857` dataset is simply one meter for floor one and two meters for floor two. This keeps things simple and easy to remember without actually using the real height of the floors, which, of course, you could do if you want to.

Our 3D routing is then able to run like any other normal 2D routing query because the data is now in 3D, thanks to our two new pgRouting functions. The query goes through, selects our data, and returns a nice GeoJSON string.

You have seen the remaining code before. It exports the results to a GeoJSON file on disk so that you can open it in QGIS for viewing. We've managed to add a couple of properties to the new GeoJSON file, including the floor number, cost in terms of distance, and the route segment type that identifies whether a segment is an indoor way or is in the form of stairs.

Calculating indoor route walk time

Our indoor routing application would not be complete without letting us know how long it would take to walk to our indoor walk now, would it? We will create a couple of small functions that you can insert into your code in the previous recipe to print out the route walk times.

How to do it...

1. Without further ado, let's take a look at some code:

```
#!/usr/bin/env python
# -*- coding: utf-8 -*-

def format_walk_time(walk_time):
    """
    takes argument: float walkTime in seconds
    returns argument: string time  "xx minutes xx seconds"
    """
```

```
    if walk_time > 0.0:
        return str(int(walk_time / 60.0)) + " minutes " +
str(int(round(walk_time % 60))) + " seconds"
    else:
        return "Walk time is less than zero! Something is
wrong"


def calc_distance_walktime(rows):
    """
    calculates distance and walk_time.
    rows must be an array of linestrings --> a route,
retrieved from the DB.
    rows[5]: type of line (stairs, elevator, etc)
    rows[3]: cost as length of segment
    returns a dict with key/value pairs route_length,
walk_time
    """

    route_length = 0
    walk_time = 0

    for row in rows:

        route_length += row[3]
        #calculate walk time
        if row[5] == 3 or row[5] == 4:  # stairs
            walk_speed = 1.2 # meters per second m/s
        elif row[5] == 5 or row[5] == 6:  # elevator
            walk_speed = 1.1  # m/s
        else:
            walk_speed = 1.39 # m/s

        walk_time += (row[3] / walk_speed)

    length_format = "%.2f" % route_length
    real_time = format_walk_time(walk_time)
    print {"route_length": length_format, "walk_time":
real_time}
```

2. Your results should show you a dictionary as follows:

    ```
    {'walk_time': '4 minutes 49 seconds', 'route_length':
    '397.19'}
    ```

    Here, it is assumed that you have placed these functions into our previous recipe and have called the function to print the results to the console.

How it works...

We have two simple functions to create walk times for our indoor routes. The first function, called `format_walk_time()`, simply takes the resulting time and converts it to a human-friendly form, showing the minutes and seconds, respectively, that are required for output.

The second function, `calc_distance_walktime()`, does the work, expecting a list object including the distance. This distance then gets summed for each route segment into a total distance value that's stored in the `route_length` variable. Our `real_time` variable is then created by calling upon the `format_walk_time` function that passes in the `walk_time` value in seconds.

Now you have a sophisticated indoor route with specified walk times for your application.

9

Topology Checking and Data Validation

In this chapter, we will cover the following topics:

- ▸ Creating a rule – only one point inside a polygon
- ▸ A point must be on the starting and ending nodes of a line only
- ▸ LineStrings must not overlap
- ▸ A LineString must not have dangles
- ▸ A polygon centroid must be within a specific distance of a line

Introduction

Topology rules allow you to enforce and test spatial relationships between different geometry sets. This chapter will build an open source set of topology rules that you can run from the command line or integrate in your python programs.

The spatial relationships described by the DE-9IM (Nine Intersect Model) are Equals, Disjoint, Intersects, Touches, Crosses, Within, Contains, and Overlaps. However, exactly how these are related is something that's unclear for most beginners. We are referring to the interior, boundary, and exterior of our geometry types: Point, LineString, and Polygon, which are used directly to perform the topology checks. These are as follows:

- ▸ **Interior**: This refers to the entire shape except for its boundary. All geometry types have interiors.
- ▸ **Boundary**: This refers to the endpoints of all linear parts of line features or the linear outline of a polygon. Only lines and polygons have boundaries.

- ▶ **Exterior**: This refers to the outside area a shape. All geometry types have exteriors.

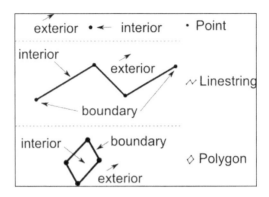

The following table summarizes the topology geometries in a more formal wording:

Geometric subtypes	Interior (I)	Boundary (B)	Exterior (E)
Point, MultiPoint	point or points	Empty set	Points not in the interior or boundary
LineString, Line	Points that are left when the boundary points are removed	Two end Points	Points not in the interior or boundary
LinearRing	All Points along the LinearRing	Empty set	Points not in the interior or boundary
MultiLineString	Points that are left when the boundary points are removed	Those Points that are in the boundaries of an odd number of its element Curves	Points not in the interior or boundary
Polygon	Points within the Rings	Set of Rings	Points not in the interior or boundary
MultiPolygon	Points within the Rings	Set of Rings of its Polygons	Points not in the interior or boundary

The definitions of the interior, boundary, and exterior of the main geometry types are described by the **Open Geospatial Consortium** (**OGC**).

In the following recipes, we will explore some custom topology rules that you could apply to any project, laying the groundwork for you to create your own set of rules.

Creating a rule – only one point inside a polygon

A long time ago in GIS history, not having more than one point present in a polygon was super important because one point per polygon was the standard way to demonstrate a topologically clean polygon with its associated attribute and ID. Today, it is still important for many other reasons, such as assigning attributes to polygons based on points inside a polygon. We must perform a spatial join between the polygon and point to assign these valuable attributes. If two points are located in one polygon, which attributes do you use? This recipe is about creating a rule to check your data beforehand to ensure that only one point is located in each polygon. If this test fails, you will get a list or errors; if it passes, the test returns `True`.

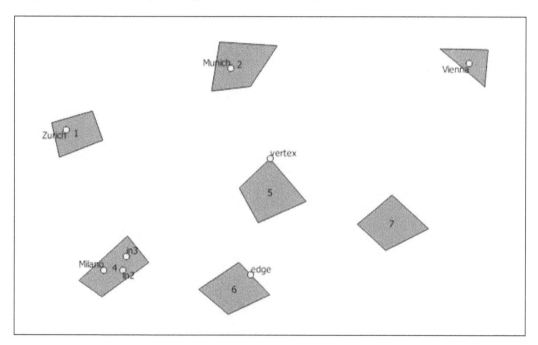

Getting ready

Data again plays the central role here, so check that your /ch09/geodata/ folder is ready with two input Shapefiles containing `topo_polys.shp` and `topo_points.shp`. The Shapely library performs the geometry topology testing. If you have followed along so far, you have it installed already; if not, install it now by referring to *Chapter 1, Setting Up Your Geospatial Python Environment*.

How to do it...

1. You will now check to see if each polygon contains a point as follows:

```python
#!/usr/bin/env python
# -*- coding: utf-8 -*-
#
# for every polygon in a polygon layer there can only be
# one point object located in each polygon
# the number of points per polygon can be defined by the
user
from utils import shp2_geojson_obj
from utils import create_shply_multigeom
import json

in_shp_poly = "../geodata/topo_polys.shp"
in_shp_point = "../geodata/topo_points.shp"

ply_geojs_obj = shp2_geojson_obj(in_shp_poly)
pt_geojs_obj = shp2_geojson_obj(in_shp_point)

shply_polys = create_shply_multigeom(ply_geojs_obj,
"MultiPolygon")
shply_points = create_shply_multigeom(pt_geojs_obj,
"MultiPoint")


def valid_point_in_poly(polys, points):
    """
    Determine if every polygon contains max one point and
that each
    point is not located on the EDGE or Vertex of the
polygon
    :param point: Point data set
    :param poly: Polygon data set
    :return: True or False if False a dictionary containing
polygon ids
    that contain no or multiple points
    """
    pts_in_polys = []
    pts_touch_plys = []

    pts_plys_geom = []
    pts_touch_geom = []
```

```
    # check each polygon for number of points inside
    for i, poly in enumerate(polys):

        pts_in_this_ply = []
        pts_touch_this_ply = []

        for pt in points:
            if poly.touches(pt):
                pts_touch_this_ply.append(
                    {'multipoint_errors_touches':
pt.__geo_interface__, 'poly_id': i,
                        'point_coord': pt.__geo_interface__})

            if poly.contains(pt):
                pts_in_this_ply.append({'multipoint_contains':
pt.__geo_interface__})

        pts_in_polys.append(len(pts_in_this_ply))
        pts_touch_plys.append(len(pts_touch_this_ply))

        # create list of point geometry errors
        pts_plys_geom.append(pts_in_this_ply)
        pts_touch_geom.append(pts_touch_this_ply)

    # identify if we have more than one point per polygon
or
    # identify if no points are inside a polygon
    no_good = dict()
    all_good = True

    # loop over list containing the number of pts per
polygon
    # each item in list is an integer representing the
number
    # of points located inside a particular polygon [4,1,0]
    # represents 4 points in polygon 1, 1 point in poly 2,
and
    # 0 points in polygon 3
    for num, res in enumerate(pts_in_polys):

        if res == 1:
            # this polygon is good and only has one point
inside
            # no points on the edge or on the vertex of
polygon
```

```
                    continue
                    # no_good['poly num ' + str(num)] = "excellen only 1
     point in poly"
            elif res > 1:
                    # we have more than one point either inside, on
     edge
                    # or vertex of a polygon
                    no_good['poly num ' + str(num)] = str(res) + "
     points in this poly"
                    all_good = False
            else:
                    # last case no points in this polygon
                    no_good['poly num ' + str(num)] = "No points in
     this poly"
                    all_good = False

        if all_good:
            return all_good
        else:
            bad_list = []
            for pt in pts_plys_geom:
                fgeom = {}
                for res in pt:
                    if 'multipoint_contains' in res:
                        hui = res['multipoint_contains']
                        print hui
                        fgeom['geom'] = hui
                bad_list.append(fgeom)
            return bad_list
            # return no_good,pts_in_polys2 # [4,0,1]


valid_res = valid_point_in_poly(shply_polys, shply_points)

final_list = []
for res in valid_res:
    if 'geom' in res:
        geom = res['geom']
        final_list.append(geom)

final_gj = {"type": "GeometryCollection", "geometries":
final_list}
print json.dumps(final_gj)
```

2. This ends the practical test using two input Shapefiles. Now for your testing pleasure, here is a simple unit test to break things down for a simple point in polygon tests. The following test code is located in the `ch09/code/ch09-01_single_pt_test_in_poly.py` file:

```python
# -*- coding: utf-8 -*-
import unittest
from shapely.geometry import Point
from shapely.geometry import Polygon

class TestPointPerPolygon(unittest.TestCase):
    def test_inside(self):

        ext = [(0, 0), (0, 2), (2, 2), (2, 0), (0, 0)]
        int = [(1, 1), (1, 1.5), (1.5, 1.5), (1.5, 1)]
        poly_with_hole = Polygon(ext, [int])

        polygon = Polygon([(0, 0), (0, 10), (10, 10),
(0, 10)])

        point_on_edge = Point(5, 10)
        point_on_vertex = Point(10, 10)
        point_inside = Point(5, 5)
        point_outside = Point(20,20)
        point_in_hole = Point(1.25, 1.25)

        self.assertTrue(polygon.touches(point_on_vertex))
        self.assertTrue(polygon.touches(point_on_edge))
        self.assertTrue(polygon.contains(point_inside))
        self.assertFalse(polygon.contains(point_outside))
        self.assertFalse(point_in_hole.within(
poly_with_hole))

if __name__ == '__main__':
    unittest.main()
```

This simple test should run nicely. If you feel like breaking it to see what happens, change the last call to the following:

```python
self.assertTrue(point_in_hole.within(poly_with_hole)
```

3. This results in the following output:

```
Failure

Traceback (most recent call last):
  File "/home/mdiener/ch09/code/ch09-01_single_pt_test_in_poly.
py", line 26, in test_inside
    self.assertTrue(point_in_hole.within(poly_with_hole))
AssertionError: False is not true
```

How it works...

We have lots of things to test to determine whether there's only one point inside the polygon. We'll start with what is defined as inside and not inside. Looking back at the introduction to this chapter, a polygon interior, exterior, and boundary can be logically defined. The position of our input points is then explicitly defined as a point that lies within a polygon, excluding points that are located on the polygon boundary, edge, or vertex. Plus, our added criterion is that only one point per polygon is allowed, thus giving errors if 0 or more points fall inside any given polygon.

Our spatial predicates include touches to find out whether the point is on the vertex or edge. If touches returns True, our point is located on the edge or vertex, which means that it is not inside. This is followed by the contains method that checks whether the point is inside our polygon. Here, we check to see that there's no more than one point inside our polygon.

The code works through importing and converting a Shapefile for processing performed by the Shapely module. As we process our polygons, we create a couple of lists to track what kind of relationship is found between them so that we can sum them up at the end, allowing us to count if zero or more than one point is inside a single polygon.

Our last bit of code then runs through a series of simple function calls, testing out the several scenarios relative to whether a point is inside the polygon or not. The final call runs through the Shapefiles with multiple polygons and points in a more realistic test. This then returns either True if no errors are found or it returns a GeoJSON printout, showing you where the errors are located.

A point must be on the starting and ending nodes of a line only

A routing network of connected edges may contain some routing logic associated with the intersections of roads that are represented as points. These points must, of course, be exactly located at the start or end of a line in order to identify these junctions. Once the junctions are found, various rules can be applied in the attributes to control your routing, for example.

A typical example would be turn restrictions that could be modeled as points:

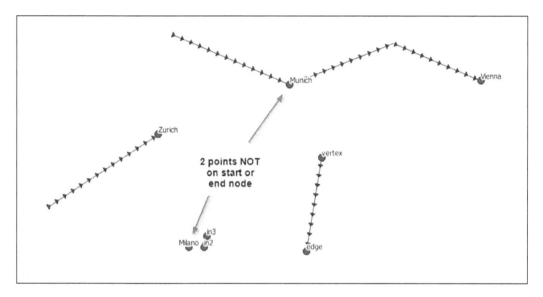

How to do it...

Our handy `utils.py` module located in the `trunk` folder helps us out with the mundane tasks of importing a Shapefile and converting it to a Shapely geometry object for us to work with.

1. Now let's create our point check code like this:

```python
#!/usr/bin/env python
# -*- coding: utf-8 -*-

from utils import shp2_geojson_obj
from utils import create_shply_multigeom
from utils import out_geoj
from shapely.geometry import Point, MultiPoint

in_shp_line = "../geodata/topo_line.shp"
in_shp_point = "../geodata/topo_points.shp"

# create our geojson like object from a Shapefile
shp1_data = shp2_geojson_obj(in_shp_line)
shp2_data = shp2_geojson_obj(in_shp_point)

# convert the geojson like object to shapely geometry
shp1_lines = create_shply_multigeom(shp1_data,
"MultiLineString")
```

```python
    shp2_points = create_shply_multigeom(shp2_data,
    "MultiPoint")


def create_start_end_pts(lines):
    '''
    Generate a list of all start annd end nodes
    :param lines: a Shapely geometry LineString
    :return: Shapely multipoint object which includes
             all the start and end nodes
    '''
    list_end_nodes = []
    list_start_nodes = []

    for line in lines:
        coords = list(line.coords)

        line_start_point = Point(coords[0])
        line_end_point = Point(coords[-1])

        list_start_nodes.append(line_start_point)
        list_end_nodes.append(line_end_point)

    all_nodes = list_end_nodes + list_start_nodes

    return MultiPoint(all_nodes)


def check_points_cover_start_end(points, lines):
    '''

    :param points: Shapely point geometries
    :param lines:Shapely linestrings
    :return:
    '''

    all_start_end_nodes = create_start_end_pts(lines)

    bad_points = []
    good_points = []
    if len(points) > 1:
        for pt in points:
            if pt.touches(all_start_end_nodes):
                print "touches"
```

```
                    if pt.disjoint(all_start_end_nodes):
                        print "disjoint" # 2 nodes
                        bad_points.append(pt)
                    if pt.equals(all_start_end_nodes):
                        print "equals"
                    if pt.within(all_start_end_nodes):
                        print "within" # all our nodes on start or
end
                    if pt.intersects(all_start_end_nodes):
                        print "intersects"
                        good_points.append(pt)
            else:
                if points.intersects(all_start_end_nodes):
                    print "intersects"
                    good_points.append(points)
                if points.disjoint(all_start_end_nodes):
                    print "disjoint"
                    good_points.append(points)


        if len(bad_points) > 1:
            print "oh no 1 or more points are NOT on a start or
end node"
            out_geoj(bad_points,
'../geodata/points_bad.geojson')
            out_geoj(good_points,
'../geodata/points_good.geojson')

        elif len(bad_points) == 1:
            print "oh no your input single point is NOT on
start or end node"

        else:
            print "super all points are located on a start or
end node" \
                "NOTE point duplicates are NOT checked"


check_points_cover_start_end(shp2_points, shp1_lines)
```

How it works...

You can attack this problem in a number of different ways. This method may not be very efficient but demonstrates how to go about solving a spatial problem.

Our logic begins with creating a function to find all the true start and end node locations of our input LineString. Shapely helps us out with some simple lists by slicing to get us the first and last coordinate pair for each of our lines. These two sets are then combined into a single list holder for all our nodes to check against.

The second function actually does the check to see whether our point is located on either the start or end node in our master list. We begin by creating the master list of start and end nodes for comparison by calling our first function. Now, if our input has more than one point, we loop through each point and check several spatial relationships. The only two that are of any real interest are disjoint and intersects. These deliver our answer by showing us which points are good and which are not.

 The within predicate could also be used instead of the intersect, but was not chosen simply because it is not always understood properly by beginners, while intersects seem to be easier to understand.

The remaining checks simply export the list of bad and good points to a GeoJSON file that you can open in QGIS to visualize.

LineStrings must not overlap

Overlapping lines are hard to find usually because you cannot see them on a map. They might be deliberate, for example, bus route network lines that might overlap. This exercise sets out to discover these overlapping lines for better or for worse.

The following diagram shows a set of two input LineStrings and you can see clearly where they overlap, but this is a cartographic visual inspection. We need this to work on many, many lines that you cannot always see as clearly.

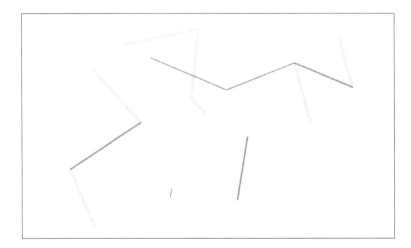

How to do it...

1. Let's dive into the code:

```python
#!/usr/bin/env python
# -*- coding: utf-8 -*-

from utils import shp2_geojson_obj
from utils import create_shply_multigeom
from utils import out_geoj

in_shp_line = "../geodata/topo_line.shp"
in_shp_overlap = "../geodata/topo_line_overlap.shp"

shp1_data = shp2_geojson_obj(in_shp_line)
shp2_data = shp2_geojson_obj(in_shp_overlap)

shp1_lines = create_shply_multigeom(shp1_data,
"MultiLineString")
shp2_lines_overlap = create_shply_multigeom(shp2_data,
"MultiLineString")

overlap_found = False

for line in shp1_lines:
    if line.equals(shp2_lines_overlap):
        print "equals"
        overlap_found = True
    if line.within(shp2_lines_overlap):
        print "within"
        overlap_found = True

# output the overlapping Linestrings
if overlap_found:
    print "now exporting overlaps to GeoJSON"
    out_int = shp1_lines.intersection(shp2_lines_overlap)
    out_geoj(out_int,
'../geodata/overlapping_lines.geojson')

    # create final Linestring only list of overlapping
lines
    # uses a pyhton list comprehension expression
    # only export the linestrings Shapely also creates  2
Points
    # where the linestrings cross and touch
```

```
      final = [feature for feature in out_int if
feature.geom_type == "LineString"]

      # code if you do not want to use a list comprehension
expresion
      # final = []
      # for f in out_int:
      #     if f.geom_type == "LineString":
      #           final.append(f)

      # export final list of geometries to GeoJSON
      out_geoj(final, '../geodata/final_overlaps.geojson')
else:
      print "hey no overlapping linestrings"
```

How it works...

Overlapping LineStrings are sometimes desirable and sometimes not. In this code, you can make some simple adjustments and have them report either situation in the form of GeoJSON. The default case is to output a GeoJSON file showing the overlapping LineStrings.

We begin the journey with the boilerplate code to convert our Shapefiles to Shapely geometries so that we can use our spatial relation predicates to filter out our overlaps. We only need two predicate equals and within to find what we are looking for. If we use intersects, these might return a false positive since both crosses() and touches() are also checked.

 We could also use the intersects predicate that is equivalent to the OR-ing of contains(), crosses(), equals(), touches(), and within() as stated in the Shapely online documentation at http://toblerity.org/shapely/manual.html#object.intersects.

A LineString must not have dangles

Dangles are like cul-de-sac (roads). You can find them only in LineStrings where a line ends and does not connect to another line segment. "To dangle in the air" refers to a LineString that is not connected to any other LineString. These are very important to identify if you are looking to ensure that a road network is connected or to identify where streets come together as they should.

A more technical description of a dangle could be described as an edge that has one or both ends that are not incidental to another edge endpoint.

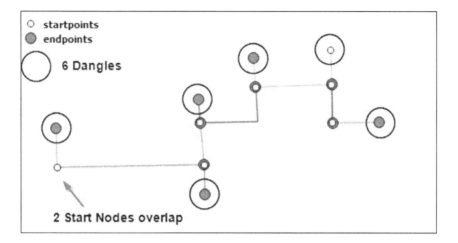

How to do it...

1. You will now check for dangles on your set of LineStrings as follows:

```python
#!/usr/bin/env python
# -*- coding: utf-8 -*-
from utils import shp2_geojson_obj
from utils import create_shply_multigeom
from utils import out_geoj
from shapely.geometry import Point

in_shp_dangles = "../geodata/topo_dangles.shp"
shp1_data = shp2_geojson_obj(in_shp_dangles)
shp1_lines = create_shply_multigeom(shp1_data,
"MultiLineString")


def find_dangles(lines):
    """
    Locate all dangles
    :param lines: list of Shapely LineStrings or
MultiLineStrings
    :return: list of dangles
    """
    list_dangles = []
    for i, line in enumerate(lines):
        # each line gets a number
```

```
                    # go through each line added first to second
                    # then second to third and so on
                    shply_lines = lines[:i] + lines[i+1:]
                    # 0 is start point and -1 is end point
                    # run through
                    for start_end in [0, -1]:
                        # convert line to point
                        node = Point(line.coords[start_end])
                        # Return True if any element of the iterable is
true.
                        # https://docs.python.org/2/library/functions.html#any
                        # python boolean evaluation comparison
                        if any(node.touches(next_line) for next_line in
shply_lines):
                            continue
                        else:
                            list_dangles.append(node)
            return list_dangles

        # convert our Shapely MultiLineString to list
        list_lines = [line for line in shp1_lines]

        # find those dangles
        result_dangles = find_dangles(list_lines)

        # return our results
        if len(result_dangles) >= 1:
            print "yes we found some dangles exporting to GeoJSON"
            out_geoj(result_dangles, '../geodata/dangles.geojson')
        else:
            print "no dangles found"
```

How it works...

Finding dangles is easy at first glance, but this is really a little more involved than one might think. So, for clarity's sake, let's explain some logic in dangle identification as pseudo code.

These are not a part of Dangle logic:

- ▶ If the start nodes of two different lines are equal, it is not a dangle
- ▶ If the end nodes of two different lines are equal, it is not a dangle
- ▶ If the start node of one line is equal to the end node of the other line, it is not a dangle
- ▶ If the end node of one line is equal to the start node of the other line, it is not a dangle

So, we need to loop over each LineString and compare the start and end points from one LineString to the next, checking if they touch each other using `touches()` from Shapely. If they do touch, we move on to the next comparison without breaking the use of `continue`. It moves to the else section and here we will catch those nice dangles and append them to the dangles list.

We are then only left with one last fun decision: to print out confirmation that we have no dangles or export the dangles to a GeoJSON fine for some visual inspection.

A polygon centroid must be within a specific distance of a line

Check that each polygon centroid is within a distance tolerance to a LineString. An example use case for such a rule could be for a routing network that defines the snap tolerance in meters from a room centroid to the nearest routing networkline. This line must be located within a certain distance; otherwise, no route can be generated, for example. The following screenshot shows the use of some dummy polygons and LineStrings, indicating the centroids that fall within our set tolerance of 20000 m in red. These are polygons that are spread far apart from Venice to Vienna:

 If you're up for some algorithm reading material, this is a nice read by Paul Bourke at `http://paulbourke.net/geometry/pointlineplane/`.

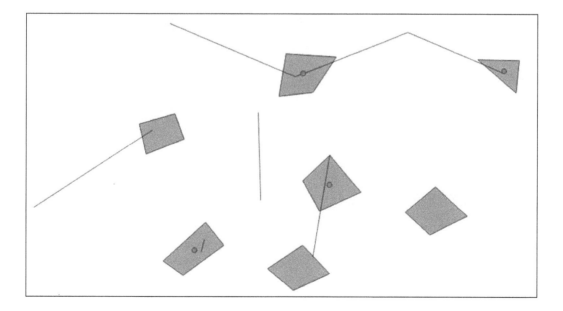

How to do it...

1. This code will now automatically find centroids outside your distance tolerance:

```python
#!/usr/bin/env python
# -*- coding: utf-8 -*-
from utils import shp2_geojson_obj
from utils import create_shply_multigeom
from utils import out_geoj

in_shp_lines = "../geodata/topo_line.shp"
shp1_data = shp2_geojson_obj(in_shp_lines)
shp1_lines = create_shply_multigeom(shp1_data,
"MultiLineString")

in_shp_poly = "../geodata/topo_polys.shp"
ply_geojs_obj = shp2_geojson_obj(in_shp_poly)
shply_polys = create_shply_multigeom(ply_geojs_obj,
"MultiPolygon")


# nearest point using linear referencing
# with interpolation and project
# pt_interpolate = line.interpolate(line.project(point))

# create point centroids from all polygons
# measure distance from centroid to nearest line segment

def within_tolerance(polygons, lines, tolerance):
    """
    Discover if all polygon centroids are within a distance
of a linestring
    data set, if not print out centroids that fall outside
tolerance
    :param polygons: list of polygons
    :param lines: list of linestrings
    :param tolerance: value of distance in meters
    :return: list of all points within tolerance
    """

    # create our centroids for each polygon
    list_centroids = [x.centroid for x in polygons]
```

```python
    # list to store all of our centroids within tolerance
    good_points = []

    for centroid in list_centroids:
        for line in lines:
            # calculate point location on line nearest to
centroid
            pt_interpolate =
line.interpolate(line.project(centroid))
            # determine distance between 2 cartesian points
            # that are less than the tolerance value in
meters
            if centroid.distance(pt_interpolate) >
tolerance:
                print "to far  " +
str(centroid.distance(pt_interpolate))
            else:
                print "hey your in  " + str(centroid.distance(pt_
interpolate))
                good_points.append(centroid)

    if len(good_points) > 1:
        return good_points
    else:
        print "sorry no centroids found within your
tolerance of " + str(tolerance)

# run our function to get a list of centroids within
tolerance
result_points = within_tolerance(shply_polys,
shp1_lines, 20000)

if result_points:
    out_geoj(result_points,
'../geodata/centroids_within_tolerance.geojson')
else:
    print "sorry cannot export GeoJSON of Nothing"
```

How it works...

Our boilerplate starter code brings in a polygon and a LineString Shapefile so that we can calculate our centroids and shortest distances. The main logic here is that we need to first create a list of centroids for each polygon, and then find the nearest point location on a line to this centroid. Of course, the last step is to get the distance between these two points in meters and check if it is less than our specified tolerance value.

Most of the comments explain the details, but the actual shortest distance to the line is accomplished using the linear referencing feature of Shapely. We have encountered this process in *Chapter 5*, *Vector Analysis*, using our snap point to a line. The `interpolate` and `project` functions do the heavy lifting to find the nearest point on the line.

This, as usual, is followed up by exporting our results to GeoJSON if any points are found with the specified tolerance value.

10
Visualizing Your Analysis

In this chapter, we will cover the following topics:

- ▶ Generating a leaflet web map with Folium
- ▶ Setting up TileStache to serve tiles
- ▶ Visualizing DEM data with Three.js
- ▶ Draping an orthophoto over a DEM

Introduction

The great part about geospatial analysis is visualization. This chapter is all about showing some ways to visualize your analysis results. Up to this point, we have used QGIS, leaflet, and Openlayers 3 to see our results. Here, we will concentrate on web mapping with some of the newest libraries to publish our data.

Most of this code will mix Python with JavaScript, HTML, and CSS.

 An awesome list of visualization techniques and libraries can be found at http://selection.datavisualization.ch/.

Generating a leaflet web map with Folium

Creating a web map with your own data is becoming easier with every new web mapping library. Folium (`http://folium.readthedocs.org/`) is a small new Python project that can create a simple web map directly from your Python code, leveraging the leaflet JavaScript mapping library. This is still more than one line, but with under 20 lines of Python code, you can have Folium generate a nice web map for you.

Getting ready

Folium requires the Jinja2 template engine alongside Pandas for data binding. The nice part about this is that both are simple to install using `pip`:

```
pip install jinja2
pip install pandas
```

Instructions on using Pandas are also found in *Chapter 1*, *Setting Up Your Geospatial Python Environment*.

How to do it...

1. Now make sure that you are in your `/ch10/code/` folder to see the live example of Folium as follows:

```python
#!/usr/bin/env python
# -*- coding: utf-8 -*-
import folium
import pandas as pd

# define the polygons
states_geojson = r'us-states.json'

# statistic data to connect to our polygons
state_unemployment = r'../www/html/US_Unemployment_Oct2012.csv'

# read the csv statistic data
state_data = pd.read_csv(state_unemployment)
```

```
# Let Folium determine the scale
map = folium.Map(location=[48, -102], zoom_start=3,
tiles="Stamen Toner")

# create the leaflet map settings
map.geo_json(geo_path=states_geojson, data=state_data,
            columns=['State', 'Unemployment'],
            threshold_scale=[5, 6, 7, 8, 9, 10],
            key_on='feature.id',
            fill_color='YlGn', fill_opacity=0.7,
            line_opacity=0.2,
            legend_name='Unemployment Rate (%)')

# output the final map file
map.create_map(path='../www/html/ch10-01_folium_map.html')
```

How it works...

Folium uses the Jinja2 Python template engine to render the final results and Pandas to bind the CSV statistic data. The code begins with importing and then defining the data sources. The GeoJSON file of the U.S. State polygons will be displayed as a **chloropleth map**. A choropleth map is one that displays data values that are classified into a defined set of data ranges, usually based on some statistical method. Within the GeoJSON data is a key-filed named id with a value the U.S. State abbreviation code. This id binds the spatial data to the statistic CSV column that also includes a corresponding id field, hence allowing us to connect our two datasets.

Folium then needs to create a map object, setting the map center coordinates alongside a zoom level and a base tile map for our background. In our case, the Stamen Toner tile set is defined.

Next up, we define the vector GeoJSON that is going to appear on top of our background map. We need to pass in the path of our source GeoJSON and the Pandas data frame object that references our CSV file columns, State and Unemployment. Next, we set the linking key value that connects our CSV with the GeoJSON data. The key_on parameter reads the id GeoJSON properties key in the feature array.

Lastly, we set the color brewer to a color we want along with the style. The legend is a D3 legend that's automatically created for us and is scaled via quantiles.

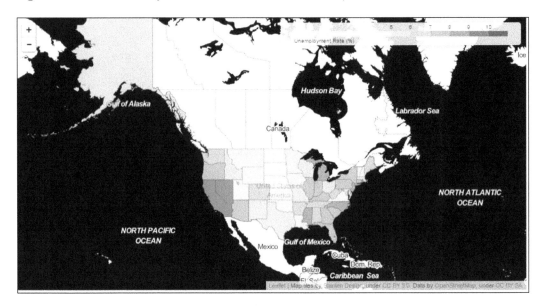

Setting up TileStache to serve tiles

Once you have data and want to get it onto the Web, a server of some sort is required. TileStache, originally developed by Michal Migurski, is a Python tile map server that can pump out vector tiles. Vector tiles are the future of web mapping and make web map applications super fast. In the end, you will have a `TileStache` instance running and serving up a simple web map.

Getting ready

A few requirements are needed to get TileStache running on your machine, including Werkzeug, PIL, SimpleJson, and Modestmaps, so we must first install these. Let's start with running our `pip install` commands like this:

> Getting `TileStache` to run on a full-blown server, such as Nginx or Apache, with `mod-python` is beyond the scope of this book but is highly recommended for production deployment (for more information on this refer to `http://modpython.org/`).

```
pip install Werkzeug
pip install modestmaps
pip install simplejson
```

The Python library called `Werkzeug` (http://werkzeug.pocoo.org/) is the WSGI server for our test application. Mapnik is not required, but go ahead and install it to view the demo application.

How to do it...

1. Now let's download the most recent code from GitHub as a ZIP from https://github.com/TileStache/TileStache/archive/master.zip.

 Use the command-line `git` if you have it installed as follows:
   `$ git clone https://github.com/TileStache/TileStache.git`

2. Unpack this into your `/ch10/TileStache-master` folder.

3. Test and check whether your installation went smoothly by going into your `/ch10/TileStache-master/` directory and entering the following command line:

   `> python tilestache-server.py -c ../tilestache.cfg`

4. After running the preceding command, you should see this:

   `* Running on http://127.0.0.1:8080/ (Press CTRL+C to quit)`

5. Now open up your web browser and type in `http://localhost:8080/`; you should see some simple text stating `TileStache belows hello`.

6. Next, try to enter `http://localhost:8080/osm/0/0/0.png`; you will get the following output:

This is the map of the world that you should be able to see.

7. To get a live scrollable map around Vancouver, British Colombia, visit `http://localhost:8080/osm/preview.html#10/49.1725/-123.0719`.

Visualizing DEM data with Three.js

You have a great 3D **Digital Elevation Model** (**DEM**) that you may want to view on a web page, so your choices are limited only to your imagination and programming skills. In this little example based on the great work of Bjorn Sandvik, we will explore the methods needed to manipulate a DEM to load a Three.js HTML-based web page.

A great plugin that I would highly recommend for QGIS is the **qgis2threejs** plugin, written by Minoru Akagi. The Python plugin code is available on GitHub at `https://github.com/minorua/Qgis2threejs` where you can find a nice `gdal2threejs.py` converter.

The resulting 3D DEM mesh can be viewed in your browser:

Getting ready

We need Jinja2 as our template engine (installed in the first section of this chapter) to create our HTML. The remaining requirements include JavaScript and our 3D DEM data. Our DEM data is from *Chapter 7, Raster Analysis*, and is located in the `/ch07/geodata/dem_3857.dem` folder, so if you have not already downloaded all the data and code, do so now.

The `gdal_translate` GDAL executable is used to convert our DEM into an ENVI `.bin` 16-bit raster. This raster will contain the elevation values that the `threejs` library can read to create the 3D mesh.

> Using an IDE is not always necessary, but in this case, the PyCharm Pro IDE is helpful since we are using HTML, JavaScript, and Python to create our results. There is also a free PyCharm community edition that I would also recommend but it lacks the HTML, JavaScript, and Jinja2 template support.

Three.js is available if you have downloaded the `/ch10/www/js` folder on your machine. If not, do so now and download the entire `/ch10/www/` folder. Inside it, you will find the folders needed for the output of HTML and the web templates used by Jinja2.

How to do it...

1. We'll start by running a subprocess call to generate the needed raster with elevation data for Three.js. Then, we'll step into the HTML template code containing a single `Jinja2` variable as follows:

```python
#!/usr/bin/env python
# -*- coding: utf-8 -*-
import subprocess

from jinja2 import Environment, FileSystemLoader


# Create our DEM

# use gdal_translate command to create an image to store
elevation values
# -scale from 0 meters to 2625 meters
#      stretch all values to full 16bit  0 to 65535
# -ot is output type = UInt16 unsigned 16bit
# -outsize is 200 x 200 px
```

```
# -of is output format ENVI raster image .bin file type
# then our input .tif with elevation
# followed by output file name .bin
subprocess.call("gdal_translate -scale 0 2625 0 65535 "
                "-ot UInt16 -outsize 200 200 -of ENVI "
                "../../ch07/geodata/dem_3857.tif "
                "../geodata/whistler2.bin")



# create our Jinja2 HTML
# create a standard Jinja2 Environment and load all files
# located in the folder templates
env = Environment(loader=FileSystemLoader(["../www/templates"]))

# define which template we want to render
template = env.get_template("base-3d-map.html")

# path and name of input 16bit raster image with our
elevation values
dem_3d = "../../geodata/whistler2.bin"

# name and location of the output HTML file we will
generate
out_html = "../www/html/ch10-03_dem3d_map.html"

# dem_file is the variable name we use in our Jinja2 HTML
template file
result = template.render(title="Threejs DEM Viewer",
dem_file=dem_3d)

# write out our template to the HTML file on disk
with open(out_html,mode="w") as f:
    f.write(result)
```

2. Our Jinja2 HTML template code only contains one simple variable called $\{\{$ dem_3d $\}\}$ so that you can see what's happening clearly:

```
#!/usr/bin/env python
<html lang="en">
<head>
    <title>DEM threejs Browser</title>
    <meta charset="utf-8">
    <meta name="viewport" content="width=device-width, user-
scalable=no, minimum-scale=1.0, maximum-scale=1.0">
```

```
    <style> body { margin: 0; overflow: hidden; }</style>
</head>
<body>
    <div id="dem-map"></div>
    <script src="../js/three.min.js"></script>
    <script src="../js/TrackballControls.js"></script>
    <script src="../js/TerrainLoader.js"></script>
    <script>

        var width  = window.innerWidth,
            height = window.innerHeight;

        var scene = new THREE.Scene();

        var axes = new THREE.AxisHelper(200);
        scene.add(axes);

        var camera = new THREE.PerspectiveCamera(45, width
        / height, 0.1, 1000);
        camera.position.set(0, -50, 50);

        var renderer = new THREE.WebGLRenderer();
        renderer.setSize(width, height);

        var terrainLoader = new THREE.TerrainLoader();
        terrainLoader.load('{{ dem_3d }}', function(data) {

            var geometry = new THREE.PlaneGeometry(60, 60,
            199, 199);

            for (var i = 0, l = geometry.vertices.length;
            i < l; i++) {
                geometry.vertices[i].z = data[i] / 65535 *
                10;
            }

            var material = new THREE.MeshPhongMaterial({
                color: 0xdddddd,
                wireframe: true
            });

            var plane = new THREE.Mesh(geometry, material);
            scene.add(plane);
```

```
        });

        var controls = new THREE.TrackballControls(camera);

        document.getElementById(
'dem-map').appendChild(renderer.domElement);

        render();

        function render() {
            controls.update();
            requestAnimationFrame(render);
            renderer.render(scene, camera);
        }

    </script>
</body>
</html>
```

How it works...

Our `gdal_translate` does the hard work for us by converting the DEM data into a raster format that Three.js can understand. The Jinja2 template HTML code shows us the required moving parts, starting with three JavaScript files. `TerrainLoader.js` reads this binary `.bin` format raster into the Three.js terrain.

Inside our HTML file, the JavaScript code shows how we can go about creating the Three.js scene where the most important part is creating `THREE.PlaneGeometry`. We assign each `geometry.vertices` the elevation height in this JavaScript `for` loop, assigning each vertex the flat plane of the elevation value.

We follow this with `MeshPhongMaterial` so that we can see the mesh on our screen as a wireframe. To view the resulting HTML file generated, you need to run a local web server and for this, Python comes with `SimpleHTTPServer` out of the box. This can be run from the command line as the following Python command:

```
> python -m SimpleHTTPServer 8080
```

Then, go visit your browser and enter `http://localhost:8080/`; select the `html` folder, and then click on the `ch10-03_dem3d_map.html` file.

Using the PyCharm IDE, you can simply open the HTML file inside PyCharm, move your mouse to the upper right-hand corner of the open file, and select a browser, such as Chrome, to open a new HTML page. PyCharm will automatically start a web server for you and display the 3D terrain in your selected browser.

Draping an orthophoto over a DEM

This time around, we are going to take our previous recipe to the next level by draping satellite imagery over our DEM to create a truly impressive 3D interactive web map.

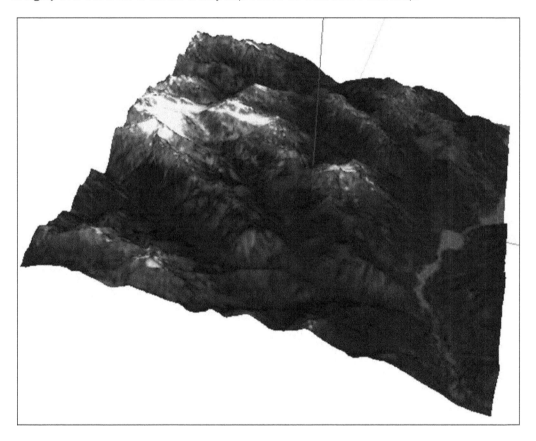

You can take a look at other orthophotos from `geogratis.ca` at `http://geogratis.gc.ca/api/en/nrcan-rncan/ess-sst/77618678-421b-4a28-a0a5-b074e5f072ff.html`.

Getting ready

To drape an orthophoto directly over our DEM, we need to make sure that the input DEM and the orthophoto have the same extent and pixel size. For this exercise, you need to complete the previous section and have data available in the `/ch10/geodata/092j02_1_1.tif` folder. This is the orthophoto that we are going to drape over the DEM.

How to do it...

1. Let's dive into some code that's full of comments for your enlightenment:

```python
#!/usr/bin/env python
# -*- coding: utf-8 -*-

import subprocess
from PIL import Image
from jinja2 import Environment, FileSystemLoader


# convert from Canada UTM http://epsg.io/3157/map    to 3857
# transform the orthophto from epsg:3157 to epsg:3857
# cut the orthophoto to same extent of DEM
subprocess.call("gdalwarp -s_srs EPSG:3157
-t_srs EPSG:3857 -overwrite "
                "-te -13664479.091 6446253.250
                -13636616.770 6489702.670"
                "/geodata/canimage_092j02_tif/092j02_1_1.tif
../geodata/whistler_ortho.tif")

# convert the new orthophoto into a 200 x 200 pixel image
subprocess.call("gdal_translate -outsize 200 200 "
                "../geodata/whistler_ortho.tif "
                "../geodata/whistler_ortho_f.tif")

# prepare to create new jpg output from .tif
processed_ortho = '../geodata/whistler_ortho_f.tif'
drape_texture = '../../geodata/whistler_ortho_f.jpg'

# export the .tif to a jpg to make is smaller for web using
pil
Image.open(processed_ortho).save(drape_texture)
```

```python
# set Jinja2 env and load folder where templates are
located
env =
Environment(loader=FileSystemLoader(["../www/templates"]))

# assign template to our HTML file with our variable inside
template = env.get_template( "base-3d-map-drape.html")

# define the original DEM file
dem_3d = "../../geodata/whistler2.bin"

# location of new HTML file to be output
out_html = "../www/html/ch10-04_dem3d_map_drape.html"

# create the new output HTML object and set variable names
result = template.render(title="Threejs DEM Drape Viewer",
dem_file=dem_3d,
                        texture_map=drape_texture)

# write the new HTML file to disk
with open(out_html,mode="w") as file:
    file.write(result)
```

2. Our Jinja2 HTML template file looks like this:

```html
<html lang="en">
<head>
    <title>DEM threejs Browser</title>
    <meta charset="utf-8">
    <meta name="viewport" content="width=device-width,
user-scalable=no, minimum-scale=1.0, maximum-scale=1.0">
    <style> body { margin: 0; overflow: hidden; }</style>
</head>
<body>
    <div id="dem-map"></div>
    <script src="../js/three.min.js"></script>
    <script src="../js/TrackballControls.js"></script>
    <script src="../js/TerrainLoader.js"></script>
    <script>

        var width  = window.innerWidth,
            height = window.innerHeight;
```

```
var scene = new THREE.Scene();
scene.add(new THREE.AmbientLight(0xeeeeee));

var axes = new THREE.AxisHelper(200);
scene.add(axes);

var camera = new THREE.PerspectiveCamera(45, width
/ height, 0.1, 1000);
camera.position.set(0, -50, 50);

var renderer = new THREE.WebGLRenderer();
renderer.setSize(width, height);

var terrainLoader = new THREE.TerrainLoader();
terrainLoader.load('{{ dem_file }}', function(data)
{

    var geometry = new THREE.PlaneGeometry(60, 60,
    199, 199);

    for (var i = 0, l = geometry.vertices.length; i
    < l; i++) {
        geometry.vertices[i].z = data[i] / 65535 *
        10;
    }

    var material = new THREE.MeshPhongMaterial({
      map: THREE.ImageUtils.loadTexture('{{
      texture_map }}')
    });

    var plane = new THREE.Mesh(geometry, material);
    scene.add(plane);

});

var controls = new THREE.TrackballControls(camera);
document.getElementById('dem-
map').appendChild(renderer.domElement);
render();
function render() {
    controls.update();
```

```
            requestAnimationFrame(render);
            renderer.render(scene, camera);
        }

    </script>
  </body>
</html>
```

How it works...

The main methodology for draping an orthophoto is the same as seen in the previous section, with a slight difference in the way we use the Three.js material rendering.

Data preparation plays the biggest and most important role once again to make things jive together. Inside our Python code, `Ch10-04_drapeOrtho.py` uses the subprocess call to execute the `gdalwarp` and `gdal_translate` command-line tools. Gdalwarp is first used by taking the original orthophoto in EPSG:3157 and converting it to the EPSG:3857 Web Mercator format. At the same time, it also cuts the original raster to the same extent as our DEM input. This extent is achieved by reading the `gdalinfo whistler.bin` raster command-line call.

After this, we need to cut the raster down to size and make a 200 x 200 pixel image to match our DEM size. This is followed by using PIL to transform the output `.tif` file into a much smaller `.jpg` file that's better suited for web presentations and speed.

With the major leg work out of the way, we can use Jinja2 to create our output HTML template and pass in two `dem_file`, variables pointing to the original DEM. The second variable called `texture_map` points to the newly created whistler `.jpg` that's used to drape over the DEM.

The final results are written to the `/ch10/www/html/ch10-04_dem3d_map_drape.html` folder for you to then open and view in the browser. To view this HTML file, you will need to start a local web server from the `/ch10/www/` directory:

```
> python -m simpleHTTPServer 8080
```

Then, visit the browser at `http://localhost.8080/` and you should see a draped image on the DEM.

11

Web Analysis with GeoDjango

In this chapter, we will cover the following topics:

- ▶ Setting up a GeoDjango web application
- ▶ Creating an indoor web routing service
- ▶ Visualizing an indoor routing service
- ▶ Creating an indoor route-type service
- ▶ Creating an indoor route from room to room

Introduction

Our final chapter is all about extending our analysis into a web application using the **Django** web framework. One of the standard Django contributed packages is known as **GeoDjango** and is found in the `django/contrib/gis` package. This is a feature-packed GIS toolset for geospatial web application development. The spatial libraries used here depend on the spatial database backend that you choose. For PostgreSQL the library requirements include GEOS, PROJ.4, and PostGIS.

Django is known for its good documentation and the `gis contrib` package installation is no exception, having its own set of instructions for you to follow at `https://docs. djangoproject.com/en/dev/ref/contrib/gis/`.

Since GeoDjango is part of the standard Django installation, you will see that your first step is to install the Django framework. For any reference on installing GeoDjango, PostgreSQL, and PostGIS, take a look at *Chapter 1, Setting Up Your Geospatial Python Environment*.

Setting up a GeoDjango web application

We need to get some basic Django groundwork done and this will be a very high-level fly over at setting up the required basics to start a Django web application. Check out the official Django tutorials for further information at `https://docs.djangoproject.com/en/dev/intro/tutorial01/`.

 If you are not familiar with Django or GeoDjango, I would highly recommend that you read through and complete the online tutorials, starting with Django at `https://docs.djangoproject.com/en/dev/` followed by the GeoDjango tutorial at `https://docs.djangoproject.com/en/dev/ref/contrib/gis/tutorial/`. For this chapter, it is assumed that you are familiar with Django, have completed the entire online Django tutorial, and are, therefore, familiar with Django concepts.

Getting ready

We are going to build a routing web service using the *Django REST framework* (`http://www.django-rest-framework.org/`). All that we need to implement is a basic web service that you can install with the help of `pip`:

```
>pip install djangorestframework==3.1.3
```

This will install version 3.1.3, the latest version. If you want to install the newest version, simply enter the following command but, beware, it might not work with this example:

```
>pip install djangorestframework
```

How to do it...

Let's now create a Django project using the `django-admin` tool as follows:

1. From the command line, enter the `/ch11/code` directory and execute this command:

    ```
    > django-admin startproject web_analysis
    ```

2. Now you will have a `/ch11/code/web_analysis/web_analysis` directory and inside it, you'll find all the standard basic Django components.

3. To create our web service, we are going to place all the services into a Django App called `api`. This app will store all our services. Creating this `api` application is as easy as typing this code:

    ```
    > cd web_analysis
    ```

Change into the newly created `web_analysis` directory:

```
> django-admin startapp api
```

Now create your new application called "api".

4. This creates a new `/ch11/code/web_analysis/api` folder and inside it you will find the default installed Django app files. Next, we need to tell Django about the Django REST Framework, GeoDjango gis app, and our new `api` application; we do this in our `/ch11/code/web_analysis/web_analysis/settings.py` file. Let's add the lines `'django.contrib.gis'`, `'rest_framework'`, and `'api'` to our `INSTALLED_APPS` variable as follows:

```
INSTALLED_APPS = (
    'django.contrib.admin',
    'django.contrib.auth',
    'django.contrib.contenttypes',
    'django.contrib.sessions',
    'django.contrib.messages',
    'django.contrib.staticfiles',

    #### GeoDjango Contrib APP
    # 'django.contrib.gis',

    #### third party apps
    'rest_framework',

    ##### our local apps
    'api',

)
```

5. To enable the GeoDjango spatial models and spatial capabilities, `'django.contrib.gis'` will allow us to access the rich geospatial framework. We have it commented out at this point since we are not going to use it until later, but feel free to uncomment it as this will do no harm. This spatial framework requires a spatial database and we will use PostgreSQL with PostGIS as our backend. Let's go ahead and change the database connection now in our `settings.py` as follows:

```
DATABASES = {
    'default': {
        # PostgreSQL with PostGIS
        'ENGINE': 'django.contrib.gis.db.backends.postgis',
```

```
                'NAME': 'py_geoan_cb', # DB name
                'USER': 'saturn', # DB user name
                'PASSWORD': 'secret', # DB user password
                'HOST': 'localhost',
                'PORT': '5432',
        }
    }
```

 The database here is referencing the same *PostgreSQL + PostGIS* database that we created earlier on in *Chapter 3, Moving Spatial Data from One Format to Another*. Visit the *Converting a Shapefile to a PostGIS table using ogr2ogr* recipe in *Chapter 3, Moving Spatial Data from One Format to Another* where we created the py_geoan_cb database, if you are going to skip ahead to this section.

6. Our final `settings.py` configuration is set up to log errors and exceptions to a log file, catching errors if any occur. First up, we'll create a new folder called `/web_analysis/logs` and add two new files called `debug.log` and `verbose.log`. We will write any errors that occur into these two files and log a request or simply print out an error to these files. So, go ahead and copy this code into the bottom of your `/web_analysis/web_analysis/settings.py` file as follows:

```
LOGGING_CONFIG = None

LOGGING = {
    'version': 1,
    'disable_existing_loggers': False,
    'formatters': {
        'verbose': {
            'format' : "[%(asctime)s] %(levelname)s
            [%(name)s:%(lineno)s] %(message)s",
            'datefmt' : "%d/%b/%Y %H:%M:%S"
        },
        'simple': {
            'format': '%(levelname)s %(message)s'
        },
    },
    'handlers': {
        'file_verbose': {
            'level': 'DEBUG',
            'class': 'logging.FileHandler',
            'filename': 'logs/verbose.log',
            'formatter': 'verbose'
        },
        'file_debug': {
```

```
                    'level': 'DEBUG',
                    'class': 'logging.FileHandler',
                    'filename': 'logs/debug.log',
                    'formatter': 'verbose'
                },
        },
        'loggers': {
            'django': {
                'handlers':['file_verbose'],
                'propagate': True,
                'level':'DEBUG',
            },
            'api': {
                'handlers': ['file_debug'],
                'propagate': True,
                'level': 'DEBUG',
            },


        }
    }

import logging.config
logging.config.dictConfig(LOGGING)
```

7. Next up, let's create a new database user and a separate PostgreSQL schema to store all our Django-related tables; otherwise, all the new Django tables will automatically be created in the PostgreSQL default schema public. Our new user is called saturn and can log in with the secret password. To create a new user, you can use the command-line tool that's run as the postgres user:

    >createuser saturn

    You can also use the PGAdmin free tool. On Ubuntu, don't forget to change to the postgres user that will allow you to create a new user on your database.

8. Now, let's create a new schema called django that will store all our Django application tables. Use PGAdmin or the SQL command to do this as follows:

    CREATE SCHEMA django AUTHORIZATION saturn;

9. With this new schema in place, we only need to assign the PostgreSQL search_path variable order to set the django schema as the first priority. To accomplish this, we need to use the SQL ALTER ROLE command as follows:

    ALTER ROLE saturn SET search_path = django, geodata, public, topology;

10. This sets the `search_path` order defining `django` as the first schema, `geodata` as the second, and so forth. This order is for all database connections for the `saturn` user. When we create our new Django tables, all of them will now automatically be created inside the `django` schema.

11. Let's go ahead now and initialize our Django project and create all the tables as follows:

```
> python manage.py migrate
```

12. The built-in Django `manage.py` command calls the `migrate` function and performs the sync in one go. Next, let's create a superuser for our application who can login and have full control of the entire web application. Then, follow the command-line instructions to enter the username, e-mail, and password as follows:

```
> python manage.py createsuperuser
```

13. With all these steps now completed, we are ready to actually get something done and build our online routing application. To test whether everything is working, run this command:

```
> python manage.py runserver 8000
```

14. Open up your local web browser and see the welcome Django default page.

Creating an indoor web routing service

Let's take all the effort we put into *Chapter 8, Network Routing Analysis*, out onto the World Wide Web. Our routing service will simply accept a starting point location, an *x*, *y* coordinate pair, a floor level, and a destination location. The indoor routing service will then calculate the shortest path and return a complete route in the form of a GeoJSON file.

Getting ready

To layout the tasks ahead, let's list out what we need to accomplish at a high level so that we're clear about where we are going:

1. Create a URL pattern to call a route service.
2. Build a view to handle an incoming URL request and deliver the appropriate GeoJSON route web response:
    1. Accept incoming request parameters.

        Start *x* coordinate.

        Start *y* coordinate.

        Start floor number.

        End *x* coordinate.

End _y_ coordinate.

End floor number.

2. Return GeoJSON LineString.

Route geometry.

Route length.

Route walk time.

We also need to let our new database user named `saturn` in order to have access to the tables located in the PostgreSQL geodata schema created in _Chapter 8, Network Routing Analysis_. Currently, only the user named `postgres` is the owner and almighty one. This needs to change so that we can keep on trucking without needing to recreate our tables as created in _Chapter 8, Network Routing Analysis_. So, let's go ahead and simply make the `saturn` user the owner of each of these tables as follows:

```
ALTER TABLE geodata.ch08_e01_networklines OWNER TO saturn;

ALTER TABLE geodata.ch08_e01_networklines_vertices_pgr OWNER TO saturn;

ALTER TABLE geodata.ch08_e02_networklines OWNER TO saturn;

ALTER TABLE geodata.ch08_e02_networklines_vertices_pgr  OWNER TO saturn;

ALTER TABLE geodata.networklines_3857 OWNER TO saturn;

ALTER TABLE geodata.networklines_3857_vertices_pgr OWNER TO saturn;
```

> If you are looking for a way to allow both the `saturn` user and any other user to gain access to these tables, you could create a PostgreSQL group role and assign the user to this role as follows:
> ```
> CREATE ROLE gis_edit   VALID UNTIL 'infinity';
> GRANT ALL ON SCHEMA geodata TO GROUP gis_edit;
> GRANT gis_edit TO saturn;
> GRANT ALL ON TABLE geodata.ch08_e01_networklines
> TO GROUP gis_edit;
> GRANT ALL ON TABLE geodata.ch08_e01_networklines_
> vertices_pgr TO GROUP gis_edit;
> GRANT ALL ON TABLE geodata.ch08_e02_networklines
> TO GROUP gis_edit;
> GRANT ALL ON TABLE geodata.ch08_e02_networklines_
> vertices_pgr TO GROUP gis_edit;
> GRANT ALL ON TABLE geodata.networklines_3857 TO
> GROUP gis_edit;
> GRANT ALL ON TABLE geodata.networklines_3857_
> vertices_pgr TO GROUP gis_edit;
> ```

Our code is now in one folder in a structure that's common to all Django web projects, so following these steps should be straightforward:

1. Let's begin by wiring up our new URL. Go ahead and open up the `urls.py` file inside your `ch11/code/web_analysis/` folder. Inside the file, you will need to enter the main URL configuration for our new web page. This file was automatically created when we created the project. Django fills in some helper text, as you can see, that shows you some basic configuration options. We need to add the `admin` app, which we will use later, and the URL for our new API. The API application will have its very own URL configuration file as you can see in the `api.urls` references, which we will create next. The `/web_analysis/urls.py` file should look like this:

```
"""web_analysis URL Configuration

The `urlpatterns` list routes URLs to views. For more
information please see:
    https://docs.djangoproject.com/en/1.8/topics/http/urls/
Examples:
Function views
    1. Add an import:  from my_app import views
    2. Add a URL to urlpatterns:  url(r'^$', views.home,
    name='home')
Class-based views
    1. Add an import:  from other_app.views import Home
    2. Add a URL to urlpatterns:  url(r'^$',
    Home.as_view(), name='home')
Including another URLconf
    1. Add an import:  from blog import urls as blog_urls
    2. Add a URL to urlpatterns:  url(r'^blog/',
    include(blog_urls))
"""
from django.conf.urls import include, url
from django.contrib import admin

urlpatterns = [
    url(r'^admin/', include(admin.site.urls)),
    url(r'^api/', include('api.urls')),
]
```

2. Next up, let's create the `/web_analysis/api/urls.py` api URLs. This file is not automatically generated so we'll create this file now. The content of this `/api/urls.py` file will be as follows:

```
from django.conf.urls import patterns, url
from rest_framework.urlpatterns import format_suffix_patterns
```

```
urlpatterns = patterns('api.views',
    #  ex valid call from to
/api/directions/
1587848.414,5879564.080,2&1588005.547,5879736.039,2
    url(r'^directions/(?P<start_coord>[-
]?\d+\.?\d+,\d+\.\d+),(?P<start_floor>\d+)&(?P<end_coord>[-
]?\d+\.?\d+,\d+\.\d+),(?P<end_floor>\d+)/$',
    'create_route', name='directions'),



)


urlpatterns = format_suffix_patterns(urlpatterns)
```

3. The regular expression looks wild as most regular expressions do. If you need some help understanding it, try referring to `https://regex101.com/#python`. Go ahead and paste this regular expression in the regular expression field:

```
(?P<start_coord>[-
]?\d+\.?\d+,\d+\.\d+),(?P<start_floor>\d+)&(?P<end_coord>[-
]?\d+\.?\d+,\d+\.\d+),(?P<end_floor>\d+)
```

4. To test your URL string, simply paste this text in the **TEST STRING** field:

```
1587848.414,5879564.080,2&1588005.547,5879736.039,2
```

5. If it's lit up in some funky colors, you are good to go:

REGULAR EXPRESSION

```
" (?P<start_coord>[-]?\d+\.?\d+,\d+\.\d+),(?P<start_floor>\d+)&(?P<end_coord>[-]?\d+\.?
\d+,\d+\.\d+),(?P<end_floor>\d+)
```

TEST STRING

```
1587848.414,5879564.080,2&1588005.547,5879736.039,2
```

Django's use of regular expressions for URL configuration is quite handy but not always obvious and explicit to read. Our URL is explained in a textual manner and would read like this:

/api/directions/start_x,start_y,start_floor&end_x,end_y,end_floor

This is a real example from your development machine. When calling the URL, it will look like this:

http://localhost:8000/api/directions/1587848.414,5879564.080,2&1588005.547,5879736.039,2

The start and end location information is separated with an & symbol, while the contents of each start parameter and end parameter are separated by a comma.

Going forward, in terms of complexity, we now need to enter the logic part of our API. Django handles this in the views. Our `/web_analysis/api/views.py` code contains the code to handle the request and response.

6. The main `def create_route` function should look familiar as it is taken directly from *Chapter 8, Network Routing Analysis,* with some modifications. A new `helper` function is created called `find_closest_network_node`. This new function is more robust and faster than our previous SQL that we used to find the node closest to any given *x, y* coordinate entered by a user:

```python
#!/usr/bin/env python
# -*- coding: utf-8 -*-

import traceback
from django.http import HttpResponseNotFound
from rest_framework.decorators import api_view
from rest_framework.response import Response
from geojson import loads, Feature, FeatureCollection
import logging
logger = logging.getLogger(__name__)
from django.db import connection


def find_closest_network_node(x_coord, y_coord, floor):
    """
    Enter a given coordinate x,y and floor number and
    find the nearest network node
    to start or end the route on
    :param x_coord: float   in epsg 3857
    :param y_coord: float   in epsg 3857
    :param floor: integer value equivalent to floor such as
     2  = 2nd floor
    :return: node id as an integer
    """
    # connect to our Database
    logger.debug("now running function
find_closest_network_node")
    cur = connection.cursor()

    # find nearest node on network within 200 m
    # and snap to nearest node
    query = """ SELECT
        verts.id as id
        FROM geodata.networklines_3857_vertices_pgr AS
verts
```

```
        INNER JOIN
          (select ST_PointFromText('POINT(%s %s %s)',
          3857)as geom) AS pt
        ON ST_DWithin(verts.the_geom, pt.geom, 200.0)
        ORDER BY ST_3DDistance(verts.the_geom, pt.geom)
        LIMIT 1;"""

    # pass 3 variables to our %s %s %s place holder in
query
    cur.execute(query, (x_coord, y_coord, floor,))

    # get the result
    query_result = cur.fetchone()

    # check if result is not empty
    if query_result is not None:
        # get first result in tuple response there is only
one
        point_on_networkline = int(query_result[0])
        return point_on_networkline
    else:
        logger.debug("query is none check tolerance value
of 200")
        return False


# use the rest_framework decorator to create our api
#  view for get, post requests
@api_view(['GET', 'POST'])
def create_route(request, start_coord, start_floor,
end_coord, end_floor):
    """
    Generate a GeoJSON indoor route passing in a start
x,y,floor
    followed by &  then the end x,y,floor
    Sample request: http:/localhost:8000/api/directions/
1587848.414,5879564.080,2&1588005.547,5879736.039,2
    :param request:
    :param start_coord: start location x,y
    :param start_floor: floor number  ex)  2
    :param end_coord: end location x,y
    :param end_floor: end floor ex)  2
    :return: GeoJSON route
    """

    if request.method == 'GET' or request.method == 'POST':
```

```python
        cur = connection.cursor()

        # parse the incoming coordinates and floor using
        # split by comma
        x_start_coord = float(start_coord.split(',')[0])
        y_start_coord = float(start_coord.split(',')[1])
        start_floor_num = int(start_floor)

        x_end_coord = float(end_coord.split(',')[0])
        y_end_coord = float(end_coord.split(',')[1])
        end_floor_num = int(end_floor)

        # use our helper function to get vertices
        # node id for start and end nodes
        start_node_id =
find_closest_network_node(x_start_coord,
                        y_start_coord,
                        start_floor_num)

        end_node_id =
find_closest_network_node(x_end_coord,
                        y_end_coord,
                        end_floor_num)

        routing_query = '''
            SELECT seq, id1 AS node, id2 AS edge,
              total_cost AS cost, layer,
              type_id, ST_AsGeoJSON(wkb_geometry) AS geoj
              FROM pgr_dijkstra(
                'SELECT ogc_fid as id, source, target,
                    st_length(wkb_geometry) AS cost,
                    layer, type_id
                 FROM geodata.networklines_3857',
                %s, %s, FALSE, FALSE
              ) AS dij_route
              JOIN  geodata.networklines_3857 AS
input_network
              ON dij_route.id2 = input_network.ogc_fid ;
          '''

        # run our shortest path query
        if start_node_id or end_node_id:
            cur.execute(routing_query, (start_node_id,
            end_node_id))
        else:
            logger.error("start or end node is None "
                        + str(start_node_id))
```

```
            return HttpResponseNotFound('<h1>Sorry NO start
            or end  node'
                            ' found within 200m</h1>')

        # get entire query results to work with
        route_segments = cur.fetchall()

        # empty list to hold each segment for our GeoJSON
          output
        route_result = []

        # loop over each segment in the result route
segments
        # create the list of our new GeoJSON
        for segment in route_segments:
            seg_cost = segment[3]      # cost value
            layer_level = segment[4]   # floor number
            seg_type = segment[5]
            geojs = segment[6]         # geojson
coordinates
            geojs_geom = loads(geojs)  # load string to
geom
            geojs_feat = Feature(geometry=geojs_geom,
                                  properties={'floor': layer_level,
                                              'length': seg_cost,
                                              'type_id':
                                              seg_type})
            route_result.append(geojs_feat)

        # using the geojson module to create our GeoJSON
Feature Collection
        geojs_fc = FeatureCollection(route_result)

        try:
            return Response(geojs_fc)
        except:
            logger.error("error exporting to json model: "+
            str(geojs_fc))
            logger.error(traceback.format_exc())
            return Response({'error': 'either no JSON or no
            key params in your JSON'})
    else:
        retun HttpResponseNotFound('<h1>Sorry not a GET or
        POST request</h1>')
```

The resulting API call has a nice web interface that's automatically generated by the **Django REST Framework** as shown in the following screenshot. The URL you need to call is also shown and should return a GeoJSON result.

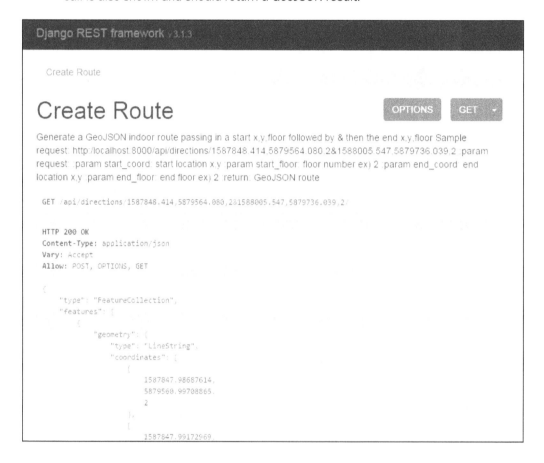

The following URL will return GeoJSON to your browser; in Chrome, it will normally just show up as simple text. IE users may download it as a file by simply opening it in Notepad++ or a local text editor to see the contents of GeoJSON:

```
http://localhost:8000/api/directions/1587848.414,5879564.080,2&158
8005.547,5879736.039,2/?format=json
```

How it works...

Our view handles the request and response using the Django REST Framework. There are two functions that do all the hard work without ever using the Django **Object Relational Mapper (ORM)**. The reason for this is two-fold: first, to show you the basics of direct Database usage without too much abstraction and the inner workings of what is going on; second, because we are using functions of PostGIS that are not available directly through the ORM of GeoDjango, such as ST_3DDistance or ST_PointFromText. We could use some of the fancy Django helpers, such as .extra(), but this would confuse everyone but an experienced Django user.

Let's discuss the first find_closest_network_node function that takes three parameters: x_coord, y_coord, and floor. The *x* and *y* coordinates should be double precision float values, while the floor is an integer. Our regular expression URL limits any request to digits so there is no need to do any extra format checking in our code.

The SQL query that finds the nearest node and returns its ID limits the search radius to 200 m, which would equal one huge room or auditorium. Then, we order by the 3D distance between the points and LIMIT the result to one since we are not routing to multiple locations.

This feeds our second function called create_route where we pass it the start coordinate, start floor integer, end coordinate, and end floor number. Our URL at /web_analysis/api/ urls.py uses a regular expression named groups that corresponds to the same names used in the request parameters of our function. This keeps things more explicit so that you know what values belong where in a query.

We begin with parsing the incoming parameters to get the exact values as floats and integers to feed our routing query. The routing query itself is unchanged from *Chapter 8, Network Routing Analysis,* so refer to this chapter for more details. The Django REST framework response sends the GeoJSON back to the client and has the ability to return it as raw text as well.

Visualizing an indoor routing service

With our wonderful API created, it's time now to visualize this indoor route returned as GeoJSON on a map. We will now dive into the Django template components to create the HTML, JS, and CSS for our front-facing web page that displays a simple slippy web map using Openlayers 3.4.0. and Bootstrap CSS.

Our new web map will display the GeoJSON on the map with a nice style alongside a menu bar where we will include later functionality.

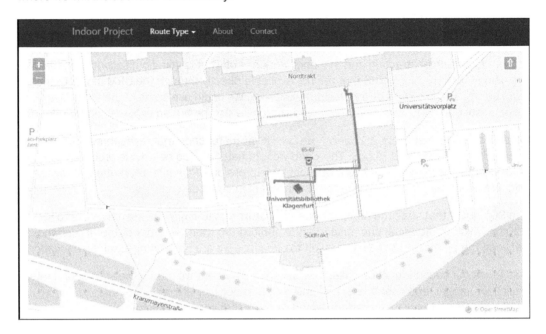

Getting ready

We need to build a few new folders and files to store new static and template content for our Django web application. Let's begin doing this by creating the /web_analysis/templates folder followed by the /web_analysis/static folder.

Inside our /static/ folder, we will place the nondynamic content of the JavaScript and CSS files. The /templates/ folder will store the HTML template files used to create our web pages.

Next up, let's tell Django /web_analysis/settings.py about the location of our new templates folder; add the os.path.join(BASE_DIR, 'templates') value to the 'DIRS' key shown here so that the TEMPLATES variable looks like this:

```
TEMPLATES = [
    {
        'BACKEND': 'django.template.backends.django.DjangoTemplates',
        'DIRS': [os.path.join(BASE_DIR, 'templates'),],
        'APP_DIRS': True,
        'OPTIONS': {
            'context_processors': [
                'django.template.context_processors.debug',
                'django.template.context_processors.request',
```

```
          'django.contrib.auth.context_processors.auth',
          'django.contrib.messages.
             context_processors.messages',
       ],
    },
  },
]
```

To manage our maps, lets create a new Django application called `maps` where we can store all our map information as follows:

> **`python manage.py startapp maps`**

Next, register your new app in the `/web_analysis/web_analysis/settings.py` `INSTALLED APPS` variable by adding this under the `api` entry `'maps'`, under the entry `'api'`.

The `/maps/urls.py` file is not automatically created so let's do this now and fill in some content as follows:

```
from django.conf.urls import patterns, url
from rest_framework.urlpatterns import format_suffix_patterns

urlpatterns = patterns('maps.views',
    #  ex valid call from to
/api/directions/1587848.414,5879564.080,2&1588005.547,5879736.039,2
    url(r'^(?P<map_name>\w+)/$', 'route_map', name='route-map'),

)

urlpatterns = format_suffix_patterns(urlpatterns)
```

We need to assign `maps/urls.py` within our main `/web_analysis/web_analysis/urls.py` so that we can freely create any URL for all our mapping needs.

Add this line to the `/web_analysis/web_analysis/urls.py` file as follows:

```
    url(r'^maps/', include('maps.urls')),
```

This means that all the URL's inside our `/maps/urls.py` will start with `http://localhost:8000/maps/`.

We are now ready to set up the static files and static contents inside `settings.py` as follows:

```
STATIC_URL = '/static/'
STATIC_FOLDER = 'static'

STATICFILES_DIRS = [
```

```
        os.path.join(BASE_DIR, STATIC_FOLDER),
]

# finds all static folders in all apps
STATICFILES_FINDERS = (
    'django.contrib.staticfiles.finders.FileSystemFinder',
    'django.contrib.staticfiles.finders.AppDirectoriesFinder',
)
```

You should have the following folders and files now in the `/static/` folder:

```
static
+---css
|       bootstrap-responsive.min.css
|       bootstrap.min.css
|       custom-layout.css
|       font-awesome.min.css
|       ol.css
|
+---img
\---js
        bootstrap.min.js
        jquery-1.11.2.min.js
        jquery.min.js
        ol340.js
```

This should be enough to set up your Django project in order for it to serve up a static map.

How to do it...

Actually serving up the map requires us to create an HTML page. We use the built-in Django template engine to build two HTML pages. The first page template is `base.html` that will hold the basics of our web map page, making it a very important part of our frontend design. What's included in this page is a set of block tags, each for separate content place holders. This allows us to quickly create new map pages based on our base template, which sets up our basic template architecture.

1. Here is the `/templates/base.html` file:

```
{% load staticfiles %}
<!DOCTYPE html>
<html lang="en">
<head>
    {% block head %}
```

```
<meta charset="utf-8">
<meta http-equiv="X-UA-Compatible" content="IE=edge">
<meta name="viewport" content="width=device-width,
initial-scale=1">
<meta name="description" content="Sample Map">
<meta name="author" content="Michael Diener">
<meta charset="UTF-8">

<title>{% block title %}Default Title{% endblock
%}</title>

<script src="{% static "js/jquery-1.11.2.min.js"
%}"></script>
<link rel="stylesheet" href="{% static
"css/bootstrap.min.css"
%}">
<script src="{% static "js/bootstrap.min.js"
%}"></script>
<link rel="stylesheet" href="{% static "css/ol.css" %}"
type="text/css">
<link rel="stylesheet" href="{% static
"css/custom-layout.css" %}" type="text/css">
<script src="{% static "js/ol340.js" %}"></script>

{% endblock head %}
</head>
<body>
{% block body %}

{% block nav %}
    <nav class="navbar navbar-inverse navbar-fixed-top">
      <div class="container">
        <div class="navbar-header">
          <button type="button" class="navbar-toggle
          collapsed" data-toggle="collapse"
          data-target="#navbar" aria-expanded="false"
          aria-controls="navbar">
            <span class="sr-only">Toggle
            navigation</span>
            <span class="icon-bar"></span>
            <span class="icon-bar"></span>
            <span class="icon-bar"></span>
          </button>
          <a class="navbar-brand" href="#">Indoor
          Project</a>
        </div>
```

```
              <div id="navbar"
              class="collapse navbar-collapse">
                <ul class="nav navbar-nav">
                  <li><a href="#about">About</a></li>
                  <li><a href="#contact">Contact</a></li>
                </ul>
              </div><!--/.nav-collapse -->
            </div>
          </nav>
      {% endblock nav %}


  {% endblock body %}
  </body>
  </html>
```

2. Now, let's move on to the actual map. A new template called `/templates/route-map.html` contains all the actual Django template blocks that are filled with HTML content as follows:

```
{% extends "base.html" %}
{% load staticfiles %}

{% block title %}Simple route map{% endblock %}

{% block body %}

{{ block.super }}

<div class="container-fluid">

    <div class="row">
      <div class="col-md-2">
        <div id="directions" class="directions">
            <form>
                <div class="radio">
                  <label>
                    <input type="radio" name="typeRoute"
                    id="routeTypeStandard" value="0"
                    checked>
                    Standard Route
                  </label>
                </div>
                <div class="radio">
                  <label>
```

```
                    <input type="radio" name="typeRoute"
                    id="routeTypeBarrierFree" value="1">
                    Barrier Free Route
                  </label>
                </div>
              <button type="submit"
              class="btn btn-default">Submit</button>
                  <br>
            </form>

        </div>
      </div>
      <div class="col-md-10">
        <div id="map" class="map"></div>
      </div>

    </div>
</div>



    <script>


          var routeUrl =
'/api/
directions/1587848.414,5879564.080,2&1588005.547,5879736.039,2&' +
sel_Val2   + '/?format=json';

          map.getLayers().push(new ol.layer.Vector({
                  source: new ol.source.GeoJSON({url:
                  routeUrl, crossDomain: true,}),
                  style:  new ol.style.Style({
                      stroke: new ol.style.Stroke({
                        color: 'blue',
                        width: 4
                      })
                    }),
                  title: "Route",
                  name: "Route"
              }));

        });
        var vectorLayer = new ol.layer.Vector({
                  source: new ol.source.GeoJSON({url:
                  geojs_url}),
```

```
                            style:  new ol.style.Style({
                                stroke: new ol.style.Stroke({
                                    color: 'red',
                                    width: 4
                                })
                            }),
                            title: "Route",
                            name: "Route"
                        });

        var map = new ol.Map({
          layers: [
            new ol.layer.Tile({
              source: new ol.source.OSM()
            }),
            vectorLayer
          ],
          target: 'map',
          controls: ol.control.defaults({
            attributionOptions: /** @type
            {olx.control.AttributionOptions} */ ({
              collapsible: false
            })
          }),
          view: new ol.View({
            center: [1587927.09817072,5879650.90059265],
            zoom: 18
          })
        });

        </script>

    {% endblock body %}
```

3. For our application to actually show these templates, we need to create a view. The view handles the request and serves `route-map.html` in return. Now, our simple view is complete:

```
from django.shortcuts import render

def route_map(request):
    return render(request, 'route-map.html')
```

How it works...

Starting with the base.html template, we set out the basic building blocks for map making. The static files and resources were set up to handle serving our JavaScript and CSS code. The base.html file is designed to allow us to add elements that are shared between multiple HTML pages such as a master page in Microsoft PowerPoint. The more blocks, that is, place holders, the better your base.

Our route-map.html contains the actual code referencing our api by calling it with a predefined, hardcoded from, to URL:

```
var geojs_url =
"http://localhost:8000/api/directions/1587898.414,5879564.080,
1&1588005.547,5879736.039,2/?format=json"
```

The /maps/views.py code is where any map logic, variables, or parameters are passed around to the template. In our code, we simply take in a request and return an HTML page. Now you have a rudimentary indoor routing service and visualization client to show off to your friends.

Creating an indoor route-type service

Building a route based on a specified type value, such as a **Barrier Free Route** or **Standard Pedestrian Route** value, is great for your users. How to build different route types is based on the available data connected to our indoor graph of ways. This example will allow a user to select the barrier-free route and our service will generate a path, avoiding obstacles such as stairs:

Getting ready

We need to access some more data on our network to allow routing types. The type of route is based on a network line type, which is stored as an attribute on each LineString. To classify our route types, we have the following lookup table schema:

Value	Route type
0	Indoor route
1	Outdoor route
2	Elevator
3	Stairs

Therefore, we want to avoid any stairs segments, which technically means avoiding `type_id = 3`.

 Optionally, you could create a lookup table to store all the possible types and their equivalent weights. These values could then be included in the calculation of the total cost value to influence the route outcome.

Now we can control how the route is generated based on certain preferences as well. A standard route search can now be set for preferences, such as taking the stairs over the elevator or vice versa, depending on your needs:

```
ALTER TABLE geodata.ch08_e01_networklines ADD COLUMN total_cost double
precision;

ALTER TABLE geodata.ch08_e02_networklines ADD COLUMN total_cost double
precision;


update geodata.networklines_3857 set total_cost = st_length(wkb_
geometry)*88
where type_id = 2;


update geodata.networklines_3857 set total_cost = st_length(wkb_
geometry)*1.8
where type_id = 3;
```

If you update `geodata.networklines_3857`, make sure the user `saturn` is the owner or has access; otherwise, your API call will break.

How to do it...

The least-cost path from any point is controlled by a basic property called `cost`. For a standard route, the cost is equal to the distance of a segment. We search for the least-cost path, which means finding the shortest path to our destination.

To control the path, we set the cost values. Creating a barrier-free route involves setting all segment types equal to `stairs` at an extraordinarily high value so that the path, that is, the distance, is huge and is, therefore, excluded in the shortest path route finding process. Our other option is to add a `WHERE` clause to the query and only accept values where `type_id` is not equal to `3`, which means that it is not of the type `stairs`. We are going to use this option in our upcoming code.

Therefore our data needs to be clean in order to allow us to assign specific costs to specific segment types in our network lines.

Now, we need to add a new parameter to capture the route type:

1. We'll update the `/api/views.py` function, `create_route()`, and add a new parameter called `route_type`. Next up is the actual query that needs to accept this new parameter. We set up a new variable called `barrierfree_q` to hold the `WHERE` clause that we will add to our original query:

```
def create_route(request, start_coord, start_floor,
end_coord, end_floor, route_type):
    base_route_q = """SELECT ogc_fid as id, source,
    target,
                    total_cost AS cost,
                    layer, type_id
                    FROM geodata.networklines_3857"""

    # set default query
    barrierfree_q = "WHERE 1=1"
    if route_type == "1":
        # exclude all networklines of type stairs
        barrierfree_q = "WHERE type_id not in (3,4)"


    routing_query = '''
        SELECT seq, id1 AS node, id2 AS edge,
        ST_Length(wkb_geometry) AS cost, layer,
        type_id, ST_AsGeoJSON(wkb_geometry) AS geoj
        FROM pgr_dijkstra('
            {normal} {type}', %s, %s, FALSE, FALSE
        ) AS dij_route
        JOIN  geodata.networklines_3857 AS input_network
        ON dij_route.id2 = input_network.ogc_fid ;
    '''.format(normal=base_route_q,
    type=barrierfree_q)
```

2. We'll update our `/api/urls.py` to input our new URL parameter, `route_type`. The newly added named group regular expression is naturally called `route_type` and only accepts numbers from 0 to 9. This then, of course, also limits you to 10 route types. So, if you want to add more types, you will need to update your `regex` as follows:

```
from django.conf.urls import patterns, url
from rest_framework.urlpatterns import format_suffix_patterns

urlpatterns = patterns('api.views',
    #   ex valid call from to
/api/directions/1587848.414,5879564.080,2&1588005.547,5879736.039,2
    url(r'^directions/(?P<start_coord>[-
]?\d+\.?\d+,\d+\.\d+),(?P<start_floor>\d+)&(?P<end_coord>[-
]?\d+\.?\d+,\d+\.\d+),(?P<end_floor>\d+)&(?P<route_type>[0-
9])/$', 'create_route', name='directions'),

)

urlpatterns = format_suffix_patterns(urlpatterns)
```

3. The `/maps/views.py` function needs a facelift too so that we can pass in the parameters. Now, it will accept `route_type` as defined in our `/api/urls.py`:

```
from django.shortcuts import render


def route_map(request, route_type = "0"):

    return render(request, 'route-map.html',
    {'route_type': route_type})
```

4. It's time to update `route-map.html` to include radio buttons that allow a user to select either a **Standard Route** or **Barrier Free Route**. The map will then update the route as soon as you click on the route type radio button:

```
{% extends "base.html" %}
{% load staticfiles %}

{% block title %}Simple route map{% endblock %}

{% block body %}

{{ block.super }}

<div class="container-fluid">

    <div class="row">
        <div class="col-md-2">
```

```html
<div id="directions" class="directions">
    <form>
        <div class="radio">
          <label>
            <input type="radio" name="typeRoute"
            id="routeTypeStandard" value="0"
            checked>
            Standard Route
          </label>
        </div>
        <div class="radio">
          <label>
            <input type="radio" name="typeRoute"
            id="routeTypeBarrierFree" value="1">
            Barrier Free Route
          </label>
        </div>
      <button type="submit"
      class="btn btn-default">Submit</button>
        <br>
    </form>

  </div>
 </div>
 <div class="col-md-10">
   <div id="map" class="map"></div>
 </div>

</div>
</div>

<script>
    var url_base = "/api/directions/";
    var start_coord = "1587848.414,5879564.080,2";
    var end_coord =  "1588005.547,5879736.039,2";
    var r_type = {{ route_type }};
    var geojs_url = url_base + start_coord + "&" +
     end_coord + "&" + sel_Val + '/?format=json';
    var sel_Val = $( "input:radio[name=typeRoute]:checked"
).val();

    $( ".radio" ).change(function() {
        map.getLayers().pop();
        var sel_Val2 = $( "input:radio[name=typeRoute]:checked"
).val();
```

```
                    var routeUrl =
    '/api/
    directions/1587848.414,5879564.080,2&1588005.547,5879736.039,2&' +
    sel_Val2  + '/?format=json';

            map.getLayers().push(new ol.layer.Vector({
                    source: new ol.source.GeoJSON({url:
                    routeUrl, crossDomain: true,}),
                    style:  new ol.style.Style({
                        stroke: new ol.style.Stroke({
                          color: 'blue',
                          width: 4
                        })
                      }),
                    title: "Route",
                    name: "Route"
                }));

        });
        var vectorLayer = new ol.layer.Vector({
                    source: new ol.source.GeoJSON({url:
                    geojs_url}),
                    style:  new ol.style.Style({
                        stroke: new ol.style.Stroke({
                          color: 'red',
                          width: 4
                        })
                      }),
                    title: "Route",
                    name: "Route"
                });

        var map = new ol.Map({
          layers: [
            new ol.layer.Tile({
              source: new ol.source.OSM()
            }),
            vectorLayer
          ],
          target: 'map',
          controls: ol.control.defaults({
            attributionOptions: /** @type
            {olx.control.AttributionOptions} */ ({
              collapsible: false
            })
```

```
        }),
        view: new ol.View({
          center: [1587927.09817072,5879650.90059265],
          zoom: 18
        })
    });

    </script>
```

{% endblock body %}

Our results for `type` = `0` or a route using stairs should look like this:

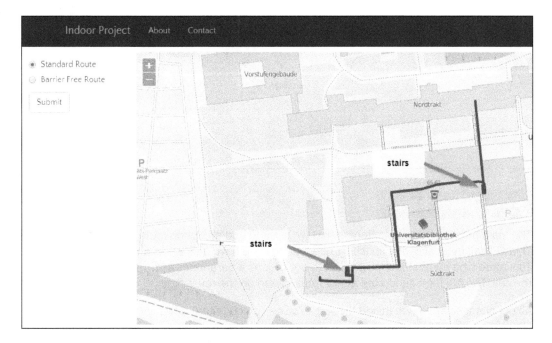

The barrier-free route will use `type` = 1, which means forced elevator use and avoiding all stairs. Your result should then look like this:

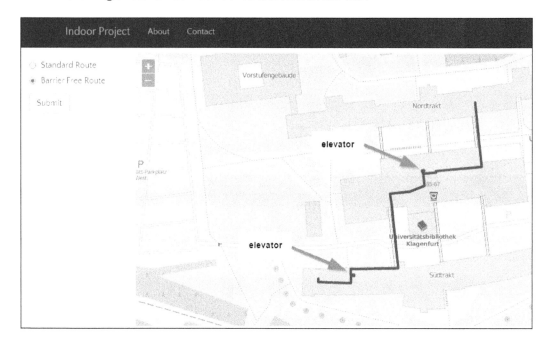

How it works...

The main part to understand here is that we need to add an option route type to our API call. This API call must accept a specific route type that we have defined as a number from 0 to 9. This route type number is then passed to our URL as a parameter and `api/views.py` runs the call. The API then generates a new route based on the route type.

All our changes are made inside the `/api/view.py` code that now includes a SQL `WHERE` clause and excludes `networklines` with a `type_id` = 3—that is, `stairs`. This query change keeps our app fast without actually increasing any Django middleware code in our views.

The frontend needs the user to select a route type with the default route type set to a standard value, such as `0`, as in the case of `stairs`. This default type is used because in most indoor environments the stairs are usually shorter. You can, of course, change this default to whatever value or criteria you'd like at any time. A radio select box is used to restrain the choice to either a standard route or a barrier-free route. Upon selecting a route type, the map automatically removes the old route and creates a new route.

Creating an indoor route from room to room

Routing from room A to room B in an indoor routing web application over multiple floors with routing types brings together all our work up to this point. We will import some room data and utilize our network to then allow a user to select a room, route from one room to the next, and select a type of route.

Getting ready

We need to import a set of room polygons for both the first and second floor as follows:

1. Import a Shapefile of the first floor room polygons as follows:

```
ogr2ogr -a_srs EPSG:3857 -lco "SCHEMA=geodata" -lco
"COLUMN_TYPES=name=varchar,room_num=integer,floor=integer"
-nlt POLYGON -nln ch11_e01_roomdata -f PostgreSQL
"PG:host=localhost port=5432 user=saturn dbname=py_geoan_cb
password=secret" e01_room_data.shp
```

2. Import a Shapefile of the second floor room polygons as follows:

```
ogr2ogr -a_srs EPSG:3857 -lco "SCHEMA=geodata" -lco
"COLUMN_TYPES=name=varchar,room_num=integer,floor=integer"
-nlt POLYGON -nln ch11_e02_roomdata -f PostgreSQL
"PG:host=localhost port=5432 user=saturn dbname=py_geoan_cb
password=secret" e02_room_data.shp
```

3. Create a new PostgreSQL view to merge all the new room data into one table so that we can query all the rooms at once:

```
CREATE OR REPLACE VIEW  geodata.search_rooms_v AS

SELECT floor, wkb_geometry, room_num FROM geodata.ch11_e01_
roomdata

UNION

SELECT floor, wkb_geometry, room_num FROM geodata.ch11_e02_
roomdata ;

ALTER TABLE geodata.search_rooms_v OWNER TO saturn;
```

How to do it...

To allow a user to route from A to B, we need to enable a route from a field and a route to a field, as follows:

1. Create a new URL to accept the new parameter of the start and end room number. The first URL for example will look like `http://localhost:8000/api/directions/10010&20043&0`, which means that the route from room number `10010` to room number `20042` using the standard route type equals to zero.

> The second URL is an extra function that you can call to only return the center coordinate of a room when you pass in the room number like this: `http://localhost:8000/directions/10010`.
>
> This function in the view *does not exist* and is left for you to do as homework.

```
    url(r'^directions/(?P<start_room_num>\d{5})&(?P<end_room_num>\
d{5})&(?P<route_type>[0-9])/$', 'route_room_to_room',
name='route-room-to-room'),
    url(r'^directions/(?P<room_num>\d{5})/$',
'get_room_centroid_node', name='room-center'),
```

2. Build a new `/api/views.py` function to find a room center coordinate and return the nearest node on `networklines` to this coordinate:

```
def get_room_centroid_node(room_number):
    '''
    Find the room center point coordinates
    and find the closest route node point
    :param room_number: integer value of room number
    :return: Closest route node to submitted room number
    '''
```

```
room_center_q = """SELECT  floor,
        ST_asGeoJSON(st_centroid(wkb_geometry))
        AS geom FROM geodata.search_rooms_v
        WHERE room_num = %s;"""

cur = connection.cursor()
cur.execute(room_center_q, (room_number,))

res = cur.fetchall()

res2 = res[0]

room_floor = res2[0]
room_geom_x = json.loads(res2[1])
room_geom_y = json.loads(res2[1])

x_coord = float(room_geom_x['coordinates'][0])
y_coord = float(room_geom_y['coordinates'][1])

room_node = find_closest_network_node(x_coord, y_coord,
room_floor)
try:
    return room_node
except:
    logger.error("error get room center " + str(room_node))
    logger.error(traceback.format_exc())
    return {'error': 'error get room center'}
```

3. Build the function inside /api/views.py to accept a start node ID, end node ID, and a route type that will then return a GeoJSON of the final route as follows:

```
def run_route(start_node_id, end_node_id, route_type):
    '''

    :param start_node_id:
    :param end_node_id:
    :param route_type:
    :return:
    '''

    cur = connection.cursor()
    base_route_q = """SELECT ogc_fid AS id, source, target,
                    total_cost AS cost,
                    layer, type_id
                    FROM geodata.networklines_3857"""
```

```
# set default query
barrierfree_q = "WHERE 1=1"
if route_type == "1":
    # exclude all networklines of type stairs
    barrierfree_q = "WHERE type_id not in (3,4)"


routing_query = '''
    SELECT seq, id1 AS node, id2 AS edge,
      ST_Length(wkb_geometry) AS cost, layer,
      type_id, ST_AsGeoJSON(wkb_geometry) AS geoj
      FROM pgr_dijkstra('
        {normal} {type}', %s, %s, FALSE, FALSE
      ) AS dij_route
      JOIN  geodata.networklines_3857 AS input_network
      ON dij_route.id2 = input_network.ogc_fid ;
  '''.format(normal=base_route_q, type=barrierfree_q)


# run our shortest path query
if start_node_id or end_node_id:
    cur.execute(routing_query, (start_node_id,
    end_node_id))
else:
    logger.error("start or end node is None "
                  + str(start_node_id))
    return HttpResponseNotFound('<h1>Sorry NO start or
    end node'
                                ' found within
                                200m</h1>')


# get entire query results to work with
route_segments = cur.fetchall()


# empty list to hold each segment for our GeoJSON
output
route_result = []


# loop over each segment in the result route segments
# create the list of our new GeoJSON
for segment in route_segments:
    seg_cost = segment[3]  # cost value
    layer_level = segment[4]  # floor number
    seg_type = segment[5]
    geojs = segment[6]  # geojson coordinates
    geojs_geom = loads(geojs)  # load string to geom
    geojs_feat = Feature(geometry=geojs_geom,
```

```
                                      properties={'floor':
                                      layer_level,
                                                  'length':
                                                  seg_cost,
                                                  'type_id':
                                                  seg_type})

        route_result.append(geojs_feat)


    # using the geojson module to create our GeoJSON
    Feature Collection
    geojs_fc = FeatureCollection(route_result)

    return geojs_fc
```

4. At last, we can create a function that our API will call to generate a response:

```
@api_view(['GET', 'POST'])
def route_room_to_room(request, start_room_num,
end_room_num, route_type):
    '''
    Generate a GeoJSON route from room number
    to room number
    :param request: GET or POST request
    :param start_room_num: an integer room number
    :param end_room_num: an integer room number
    :param route_type: an integer room type
    :return: a GeoJSON linestring of the route
    '''

    if request.method == 'GET' or request.method == 'POST':

        start_room = int(start_room_num)
        end_room = int(end_room_num)

        start_node_id = get_room_centroid_node(start_room)
        end_node_id = get_room_centroid_node(end_room)

        res = run_route(start_node_id, end_node_id,
        route_type)

        try:
            return Response(res)
        except:
            logger.error("error exporting to json model: "
            + str(res))
            logger.error(traceback.format_exc())
```

```
                    return Response({'error': 'either no JSON or no
                    key params in your JSON'})
            else:
                return HttpResponseNotFound('<h1>Sorry not a GET or
                POST request</h1>')
```

5. Add a URL to `/api/urls.py` to access a list of all available rooms:

```
url(r'^rooms/$', 'room_list', name='room-list'),
```

6. Create an API service to return a JSON array of all room numbers. This array is used in autocomplete fields, `route-from`, and `route-to`. We use the Twitter `Typeahead.js` JavaScript library to handle our autocomplete dropdown type hinting. As a user, all you need to do is type 1, for example, and all the rooms beginning with 1 will show up as `10010` (check this out at `http://twitter.github.io/typeahead.js/examples/`):

```
@api_view(['GET', 'POST'])
def room_list(request):
    '''

    http://localhost:8000/api/rooms
    :param request: no parameters GET or POST
    :return: JSON Array of room numbers
    '''
    cur = connection.cursor()
    if request.method == 'GET' or request.method == 'POST':

        room_query = """SELECT room_num FROM
        geodata.search_rooms_v"""

        cur.execute(room_query)
        room_nums = cur.fetchall()

        room_num_list = []
        for x in room_nums:
            v = x[0]
            room_num_list.append(v)

        try:
            return Response(room_num_list)
        except:
            logger.error("error exporting to json model: "
            + str(room_num_list))
            logger.error(traceback.format_exc())
            return Response({'error': 'either no JSON or no
            key params in your JSON'})
```

7. Our final `base.html` template is complete, containing all the spice needed for our final route from room to room as follows:

```
{% load staticfiles %}
<!DOCTYPE html>
<html lang="en">
<head>
    {% block head %}

    <meta charset="utf-8">
    <meta http-equiv="X-UA-Compatible" content="IE=edge">
    <meta name="viewport" content="width=device-width,
    initial-scale=1">
    <meta name="description" content="Sample Map">
    <meta name="author" content="Michael Diener">
    <meta charset="UTF-8">

    <title>{% block title %}Default Title{% endblock
    %}</title>

    <script src="{% static "js/jquery-1.11.2.min.js"
    %}"></script>
    <link rel="stylesheet" href="{% static
    "css/bootstrap.min.css" %}">
    <script src="{% static "js/bootstrap.min.js"
    %}"></script>
    <link rel="stylesheet" href="{% static "css/ol.css" %}"
    type="text/css">
    <link rel="stylesheet" href="{% static
    "css/custom-layout.css" %}" type="text/css">
    <script src="{% static "js/ol340.js" %}"></script>

    {% endblock head %}
</head>
<body>
{% block body %}

    {% block nav %}
        <nav
        class="navbar navbar-inverse navbar-fixed-top">
          <div class="container">
            <div class="navbar-header">
              <button type="button"
              class="navbar-toggle collapsed"
              data-toggle="collapse" data-target="#navbar"
              aria-expanded="false" aria-controls="navbar">
```

```
                    <span class="sr-only">Toggle
                    navigation</span>
                    <span class="icon-bar"></span>
                    <span class="icon-bar"></span>
                    <span class="icon-bar"></span>
                </button>
                <a class="navbar-brand"
                href="http://www.indrz.com"
                target="_blank">Indoor Project</a>
            </div>
            <div id="navbar"
            class="collapse navbar-collapse">
                <ul class="nav navbar-nav">
                    <li><a href="#about"
                    target="_blank">About</a></li>
                    <li><a href="https://github.com/mdiener21/"
                    target="_blank">Contact</a></li>
                </ul>
            </div><!--/.nav-collapse -->
        </div>
    </nav>
{% endblock nav %}


{% endblock body %}
</body>
</html>
```

8. Now we'll create our final `route-map.html` template and the JavaScript that goes with it as follows:

```
{% extends "base.html" %}
{% load staticfiles %}

{% block title %}Simple route map{% endblock %}

{% block head %}
{{ block.super }}
    <script src="{% static "js/bloodhound.min.js"
    %}"></script>
    <script src="{% static "js/typeahead.bundle.min.js"
    %}"></script>
{% endblock head %}

{% block body %}
```

```
{{ block.super }}

<div class="container-fluid">

    <div class="row">
      <div class="col-md-2">
        <div id="directions" class="directions">
            <form id="submitForm">
              <div id="rooms-prefetch" class="form-group">
                <label for="route-to">Route From:</label>
                <input type="text"
                class="typeahead form-control"
                id="route-to" placeholder="Enter Room
                Number">
              </div>
              <div id="rooms-prefetch" class="form-group">
                <label for="route-from">Route To:</label>
                <input type="text"
                class="typeahead form-control"
                id="route-from" placeholder="Enter Room
                Number">
              </div>

                <div class="radio">
                  <label>
                    <input type="radio" name="typeRoute"
                    id="routeTypeStandard" value="0"
                    checked>
                    Standard Route
                  </label>
                </div>
                <div class="radio">
                  <label>
                    <input type="radio" name="typeRoute"
                    id="routeTypeBarrierFree" value="1">
                    Barrier Free Route
                  </label>
                </div>
              <button id="enterRoute" type="submit"
              class="btn btn-default">Go !</button>
                <br>
            </form>

        </div>
      </div>
```

```html
        <div class="col-md-10">
          <div id="map" class="map"></div>
        </div>

      </div>
    </div>

    <script>  {% include 'routing.js' %} </script>


    <script>
        var roomNums = new Bloodhound({
          datumTokenizer: Bloodhound.tokenizers.whitespace,
          queryTokenizer: Bloodhound.tokenizers.whitespace,
          prefetch:
'http://localhost:8000/api/rooms/?format=json'
        });

        // passing in `null` for the `options` arguments will
        result in the default
        // options being used
        $('#rooms-prefetch .typeahead').typeahead(null, {
          name: 'countries',
            limit: 100,
          source: roomNums
        });


        $( "#submitForm" ).submit(function( event ) {
        {#  alert( "Handler for .submit() called."  );#}
            var startNum = $('#route-from').val();
            var endNum = $('#route-to').val();
            var rType = $( "input:radio[name=typeRoute]:checked"
).val();
             addRoute(startNum, endNum, rType);
          event.preventDefault();
        });


    </script>


{% endblock body %}
```

9. Our `maps/templates/routing.js` contains the functions needed to call the routing API as follows:

```
var url_base = "/api/directions/";
var start_coord = "1587848.414,5879564.080,2";
var end_coord =  "1588005.547,5879736.039,2";
var sel_Val = $( "input:radio[name=typeRoute]:checked"
).val();
var geojs_url = url_base + start_coord + "&" +
end_coord + "&" + sel_Val + '/?format=json';

// uncomment this code if you want to reactivate
// the quick static demo switcher
//$( ".radio" ).change(function() {
//    map.getLayers().pop();
//    var sel_Val2 = $( "input:radio[name=typeRoute]:check
ed" ).val();
//    var routeUrl = '/api/
directions/1587848.414,5879564.080,2&1588005.547,5879736.039,2&' +
sel_Val2  + '/?format=json';
//
//    map.getLayers().push(new ol.layer.Vector({
//          source: new ol.source.GeoJSON({url:
routeUrl}),
//          style:  new ol.style.Style({
//              stroke: new ol.style.Stroke({
//                  color: 'blue',
//                  width: 4
//                })
//            }),
//          title: "Route",
//          name: "Route"
//       }));
//
//});

var vectorLayer = new ol.layer.Vector({
          source: new ol.source.GeoJSON({url:
          geojs_url}),
          style:  new ol.style.Style({
              stroke: new ol.style.Stroke({
                color: 'red',
                width: 4
              })
            }),
          title: "Route",
```

```
                name: "Route"
            });

    var map = new ol.Map({
      layers: [
        new ol.layer.Tile({
          source: new ol.source.OSM()
        })

          ,
        vectorLayer
      ],
      target: 'map',
      controls: ol.control.defaults({
        attributionOptions: /** @type
        {olx.control.AttributionOptions} */ ({
          collapsible: false
        })
      }),
      view: new ol.View({
        center: [1587927.09817072,5879650.90059265],
        zoom: 18
      })
    });

function addRoute(fromNumber, toNumber, routeType) {
    map.getLayers().pop();
    console.log("addRoute big"+ String(fromNumber));
    var baseUrl = 'http://localhost:8000/api/directions/';
    var geoJsonUrl = baseUrl + fromNumber + '&' + toNumber
    + '&' + routeType +'/?format=json';
    console.log("final url " + geoJsonUrl);
    map.getLayers().push(new ol.layer.Vector({
                source: new ol.source.GeoJSON({url:
                geoJsonUrl}),
                style:  new ol.style.Style({
                    stroke: new ol.style.Stroke({
                      color: 'purple',
                      width: 4
                    })
                  }),
                title: "Route",
                name: "Route"
            }));
    }
```

10. Now, go ahead and enter 1 and see the autocomplete in action; then, select **Route To:** and enter 2 to see the second floor options. Finally, click on **GO!** and see the magic happen:

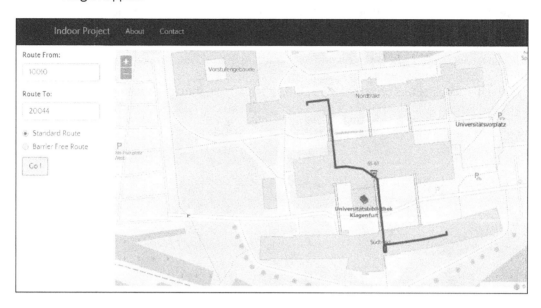

How it works...

The internal workings of each step are shown, so let's go through all of them one at a time. We start off with the data import of our new room dataset as the basic starting and ending points of our new indoor routing tool. We lay out the top-level structure of our API with some new URLs, defining how we will call the new routing service, and then define these variables. Our regular expressions handle the correct data types that are passed in the URL without any exceptions.

These URL patterns are then used by api/views.py to actually accept the incoming room numbers and route types to generate our new route. This generation is split up into a few functions to increase usability. The get_room_centroid_node() function is necessary so that we can find the middle point of the room and then find the next nearest node on a network. We could also simply use the polygon geometry to find the nearest node but this can lead to ambiguity if the rooms are large and the entrances are close to each other. The centroid method is much more reliable and does not add too much overhead.

The run_route() function actually then runs find_closest_network_node(), which we created earlier, making things work well together. The run_route function then generates our GeoJSON result as it is passed in the start node ID, end node ID, and the route type.

The `route_room_to_room()` function is small as the heavy lifting has already been completed by our other functions. It simply inputs the URL parameters called by our API call, as seen in `http://localhost:8000/api/directions/10010&20043&0`. The final steps after step 6 are for the user interface. We need to provide the user with a list of rooms that are available to route from and to. The `/api/rooms` URL provides exactly this, delivering a JSON array of room numbers. The input fields are bootstrap inputs with Twitter `Typeahead.js` and `Bloodhound.js` to prefetch remote data. As a user, you simply enter a number and, bingo, a list appears. More detailed instructions on the JavaScript side of things are a little beyond the scope of this book, but these are, thankfully, kept to a minimum.

All in all, you now have a fully functional indoor mapping web application with a basic set of indoor 3D routing functions that you can expand on at any time.

Other Geospatial Python Libraries

We have covered many libraries and examples but we haven't covered them all. This appendix is meant to quickly go over the other libraries out there that play a special role in the Python geospatial working environment. This list is definitely not complete and I have not had the pleasure of working with all these libraries at the time of writing.

The listing is a resource for further reading and experiments that will, hopefully, provide you with a step in the right direction to solve your specific problems. Each description of a library starts with the official library name followed by a short description and a link to the web page:

Library name	Description	Website
Rtree	This is a Python wrapper of `libspatialindex` that provides advanced spatial indexing features	`http://toblerity.org/rtree`
rasterio	This is a Mapbox creation that aims at working with rasters in an easier manner	`https://github.com/mapbox/rasterio`
Fiona	This focuses on reading and writing data in the standard Python I/O style	`http://toblerity.org/fiona`
geopy	This helps geocoding in Python	`http://www.geopy.org`
PyQGIS	This is the Python interface to QGIS (formerly known as Quantum GIS) that helps extend QGIS and more	`http://pythongisbook.com`
GeoPandas	This is an extension of the pandas library and handles geospatial database	`http://geopandas.org/`
MapFish	This is Python's geospatial web framework	`http://mapfish.org`
PyWPS	This client interacts with various open geospatial standard services	`http://pywps.wald.intevation.org`
pycsw	This provides a metadata catalog interface	`http://pycsw.org`

Library name	Description	Website
GeoNode	This provides Python geospatial content management for the Web and is built on the Django web framework and GeoServer	`http://geonode.org`
mapnik	This is a map visualization library to create maps for web tile cache	`http://mapnik.org`
cartopy	This is mapping made easy in Python-shapely	`http://scitools.org.uk/cartopy`
Kartograph	This creates SVG maps or web maps	`http://kartograph.org`
basemap	This is an extension of matplotlib in combination with descartes	`http://matplotlib.org/basemap`
SciPy	This is a collection of Python libraries for scientific data analysis that are bundled or available as individual installations	`http://www.scipy.org`
GeoAlchemy	This is a spatial extension to SQLAlchemy that works with the spatial database PostGIS	`http://geoalchemy.org`
pyspatialite	This helps you work with spatialite databases of geospatial data	`https://pypi.python.org/pypi/pyspatialite`
gpxpy	This helps when working with GPS data in the standard GPX format in a Python - friendly format	`http://www.trackprofiler.com/gpxpy/index.html`
ShaPy	This is a pure Python version of Shapely with no dependencies	`https://github.com/karimbahgat/Shapy`
pyshp	This reads and writes Shapefiles in pure Python	`https://github.com/GeospatialPython/pyshp`
TileCache	This is an implementation of a WMS-C (catalog) **Tile Mapping Server** (**TMS**) server	`http://tilecache.org`
TileStache	This is a Python-based server application that can serve up map tiles based on rendered geographic data	`http://www.tilestache.org`
FeatureServer	This is a restful feature service to easily get, edit, delete, and update features over the Web with the help of HTTP	`http://featureserver.org`
GeoScript	This is an implementation of Python, giving spatial analysis functionality to other scripting languages and Python is one of them; it is similar to Shapely	`http://www.geoscript.org`
karta	This is a Leatherman for geographic analyses	`http://ironicmtn.com/karta`

B

Mapping Icon Libraries

Finding the perfect mapping icon set is hard. The following list provides some of the better map symbols around, for your web map application:

Library name	Description	Website
map-icons	This is an icon font that's used with the Google Maps API and Google Places API using SVG markers and icon labels	`http://map-icons.com/`
Maki	This creates Mapbox Pixel-perfect icons for web cartography	`https://www.mapbox.com/maki`
map icons	This focuses on reading and writing data in the standard Python IO style	`https://mapicons.mapsmarker.com/`
Integration and Application Network	This creates 2782 custom vector symbols	`http://ian.umces.edu/symbols/`
OSM icons	This is a set of free SVG icons that are used for OSM maps	`http://osm-icons.org/wiki/Icons`
OSGeo map symbol set	This is a collection of links to map icons	`http://wiki.osgeo.org/wiki/OSGeo_map_symbol_set`
SJJB collection	This is a set of PD/CC0 SVG map icons and tools to generate PNG icons	`https://github.com/twain47/Open-SVG-Map-Icons` and `http://www.sjjb.co.uk/mapicons/contactsheet`

Library name	Description	Website
OSM map-icons	This creates an OpenStreetMap set of icons	`https://github.com/openstreetmap/map-icons/tree/master/svg`
opensreetmap-carto	Andy Allan created these set of mapping icons in PNG	`https://github.com/gravitystorm/openstreetmap-carto/tree/master/symbols`

Index

Thank you for buying
Python Geospatial Analysis Cookbook

About Packt Publishing

Packt, pronounced 'packed', published its first book, *Mastering phpMyAdmin for Effective MySQL Management*, in April 2004, and subsequently continued to specialize in publishing highly focused books on specific technologies and solutions.

Our books and publications share the experiences of your fellow IT professionals in adapting and customizing today's systems, applications, and frameworks. Our solution-based books give you the knowledge and power to customize the software and technologies you're using to get the job done. Packt books are more specific and less general than the IT books you have seen in the past. Our unique business model allows us to bring you more focused information, giving you more of what you need to know, and less of what you don't.

Packt is a modern yet unique publishing company that focuses on producing quality, cutting-edge books for communities of developers, administrators, and newbies alike. For more information, please visit our website at www.packtpub.com.

About Packt Open Source

In 2010, Packt launched two new brands, Packt Open Source and Packt Enterprise, in order to continue its focus on specialization. This book is part of the Packt open source brand, home to books published on software built around open source licenses, and offering information to anybody from advanced developers to budding web designers. The Open Source brand also runs Packt's open source Royalty Scheme, by which Packt gives a royalty to each open source project about whose software a book is sold.

Writing for Packt

We welcome all inquiries from people who are interested in authoring. Book proposals should be sent to author@packtpub.com. If your book idea is still at an early stage and you would like to discuss it first before writing a formal book proposal, then please contact us; one of our commissioning editors will get in touch with you.

We're not just looking for published authors; if you have strong technical skills but no writing experience, our experienced editors can help you develop a writing career, or simply get some additional reward for your expertise.

QGIS Python Programming Cookbook

ISBN: 978-1-78398-498-5   Paperback: 340 pages

Over 140 recipes to help you turn QGIS from a desktop GIS tool into a powerful automated geospatial framework

1. Use Python and QGIS to create and transform data, produce appealing GIS visualizations, and build complex map layouts.

2. Learn undocumented features of the new QGIS processing module.

3. A set of user-friendly recipes that can automate the entire geospatial workflows by connecting Python GIS building blocks into comprehensive processes.

Learning R for Geospatial Analysis

ISBN: 978-1-78398-436-7   Paperback: 364 pages

Leverage the power of R to elegantly manage crucial geospatial analysis tasks

1. Write powerful R scripts to manipulate your spatial data.

2. Gain insight from spatial patterns utilizing R's advanced computation and visualization capabilities.

3. Work within a single spatial analysis environment from start to finish.

Learning Geospatial Analysis with Python

ISBN: 978-1-78328-113-8   Paperback: 364 pages

Master GIS and Remote Sensing analysis using Python with these easy to follow tutorials

1. Construct applications for GIS development by exploiting Python.

2. Focuses on built-in Python modules and libraries compatible with the Python Packaging Index distribution system – no compiling of C libraries necessary.

3. This is a practical, hands-on tutorial that teaches you all about Geospatial analysis in Python.

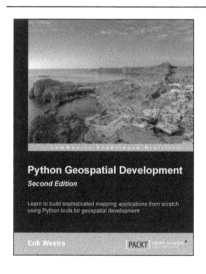

Python Geospatial Development

Second Edition

ISBN: 978-1-78216-152-3   Paperback: 508 pages

Learn to build sophisticated mapping applications from scratch using Python tools for geospatial development

1. Build your own complete and sophisticated mapping applications in Python.

2. Walks you through the process of building your own online system for viewing and editing geospatial data.

3. Practical, hands-on tutorial that teaches you all about geospatial development in Python.

Please check **www.PacktPub.com** for information on our titles

www.ingramcontent.com/pod-product-compliance
Lightning Source LLC
Chambersburg PA
CBHW062109050326
40690CB00016B/3263